CAREC Road Safety Engineering Manual 6

IDENTIFYING, INVESTIGATING, AND TREATING BLACKSPOTS

MAY 2024

Contents

Tables, Figures, and Boxes

TABLES

FIGURES

BOXES

Abbreviations

ADB	–	Asian Development Bank
BAC	–	blood alcohol concentration
BCR	–	benefit–cost ratio
CAREC	–	Central Asia Regional Economic Cooperation
CRF	–	crash reduction factor
DCC	–	description for classifying crashes
FSI	–	fatal and serious injury
GDP	–	gross domestic product
km	–	kilometer
km/h	–	kilometer per hour
m	–	meter
mm	–	millimeter
NGO	–	nongovernment organization
POS	–	pedestrian operated signal
ROR	–	run-off-road
RSA	–	Road Safety Audit
RSE	–	Road Safety Engineering

Purpose of This Manual

Ministers of all Central Asia Regional Economic Cooperation (CAREC) countries have endorsed the CAREC Road Safety Strategy, 2017–2030, which supports and encourages governments and road authorities to plan, design, construct, and maintain roads with road safety as a key and specific objective.

Although the role of human error in road crashes is considerable, road infrastructure has a key role in reducing the likelihood of a crash. When a crash happens, the road infrastructure plays a most significant role in minimizing the consequences of the crash.

Manual 6 (Identifying, Investigating, and Treating Blackspots) in the CAREC series of road safety engineering manuals is a practical point of reference for removing high-frequency crash locations from roads in CAREC countries. The manual is for use by highway and road safety engineers, project managers, planners, municipal engineers, traffic police, and consultants and representatives of design institutes and road agencies. It focuses on finding and treating the road infrastructure features that can contribute to a crash or increase the consequences of a crash.

It provides an essential understanding of:

- the role of engineers in providing safe roads,
- the chain of events that lead to a crash,
- the Safe System approach to safety,
- the need for, and benefits of, accurate crash data,
- identifying blackspots,
- investigating a blackspot, including the tools available to find crash patterns,
- treating blackspots with effective countermeasures, and
- determining the priority for funding blackspot treatments.

The manual offers full information about the blackspot process for those who are responsible for identifying and investigating these locations (practitioners), as well as full information about establishing a national (or municipal) blackspot removal program for those who manage the road network (policymakers). It encourages road authorities and traffic police to collaborate by sharing crash data, the cornerstone of a national blackspot removal program.

The manual also encourages road authorities across the CAREC region to put more resources into removing blackspots. Experience shows that for every $1 spent on treating blackspots, $4 in crash savings returns to the community. Therefore, a blackspot removal program is a prudent investment in road safety.

This manual was prepared under a grant offered by the Asian Development Bank (ADB) as Technical Assistance 6591 REG: Enhancing Road Safety for Central Asia Regional Economic Cooperation Member Countries (Phase 2). The production of this manual was administered and managed by the CAREC Secretariat at ADB. The Secretariat team includes Ritu Mishra and Pilar Sahilan. The author of this manual is Phillip Jordan, ADB road safety engineering consultant.

I. Engineers Have an Important Role in Road Safety

1.   Treating high-crash frequency locations with a package of countermeasures is only one of many elements in a successful road safety strategy, but it is a proven and cost-effective element.

2.   This manual outlines the practical aspects of the blackspot investigation process, whether for an individual crash site or many sites in a national or municipal blackspot removal program. There are three key stages in the blackspot process:

- identifying a site as a blackspot;
- investigating the site, seeking contributing factors to the crashes, and reporting the benefit–cost ratio (BCR) of the recommended package of countermeasures for funding approval; and
- installing cost-effective treatments at the site and monitoring their effectiveness.

3.   This manual details these three stages and encourages Central Asia Regional Economic Cooperation (CAREC) road authorities, traffic police, and other stakeholders to coordinate their activities in support of national blackspot removal programs.

4.   Successful blackspot removal programs need an experienced local road safety engineering profession as a starting point. While importing experts from other parts of the world is possible, a sustainable blackspot removal program that will provide positive returns for years ahead needs a local road safety engineering section in the national road authority. Road safety engineering is an important job, requiring qualified and experienced professionals to ensure the safety needs of all road users are met. Road safety engineers are the specialists who identify, investigate, and treat blackspots under a national blackspot removal program. Their work adds value to the ongoing work of a road authority.

5.   Road safety engineering is an increasingly important profession as global efforts in road safety increase. In the past, it was common for crashes to be blamed on the driver, rider, or pedestrian; it was easy to blame them for speeding or using their phone while driving. No one questioned the road infrastructure, with its outdated geometry (unsafe for today's volumes and speeds), lack of signs/delineation, or unsafe roadsides. Deflecting the blame onto the road user was easier than accepting that the road itself had a role in safety.

6.   But times are changing. It is now widely recognized that the road and the roadside contribute to a large proportion of crashes. It is no longer acceptable to place all blame for crashes on the driver, rider, or pedestrian. The Safe System approach (see Chapter 1, Section C) expects all groups to be responsible for road safety. It expects road users to comply with all road rules and for the road authority to be responsible for the management of safe roads. To do this well, road authorities need teams of experienced road safety engineers.

Engineers at work. Road safety engineers have a responsibility and role in national blackspot removal programs.

7.   One of the key messages from this manual is for decision-makers in road agencies across the CAREC region to continue prioritizing the road safety needs of its road users and investing in cost-effective, proven road safety programs. This will require an investment in human resources, in the form of expanding teams of road safety engineers. Additionally, it will require the development of national blackspot removal programs. These programs have a proven track record in enhancing road safety, and their adoption by road authorities across the CAREC region is strongly recommended.

A.   What is a road crash?

8.   It is common for the public to refer to road accidents during conversations about recent newsworthy incidents on the roads. They do this without giving much thought to the term "accident," as they have always used that term. But accident implies that no one could have prevented the event, and nothing could have been done to avoid it. For these reasons, the road safety profession now prefers to use the term crash (or collision) instead of accident.

9.   Therefore, in this manual, the term "crash" is used to emphasize that these tragic impacts are preventable and that they are not inevitable events beyond human control. Road safety professionals know that crashes are avoidable and preventable if enough resources and attention are applied to the subject.

10.   While most people understand the consequences of a crash, it is not easy to offer a definition for a crash that covers every situation. Must it always lead to a fatality? Must it always involve two motor vehicles? There are so many varied combinations of vehicles, road users, and crash types that defining them is difficult.

11.   One widely accepted definition for a road crash states: "A crash is a rare, random, multifactorial event in which one or more road users have failed to cope with their environment."[1]

12.   Crashes are rare, but they do occur. They are said to be random as they happen at different places and different times and involve different road users. No one can predict where or when the next serious casualty crash will occur in their country. Crashes have many contributing factors and are, therefore, said to be multifactorial events. And crashes invariably involve road users (drivers, passengers, bicyclists, pedestrians, and others) who have made a mistake or perhaps several mistakes.

"A crash is a rare, random, multifactorial event in which one or more road users have failed to cope with their environment"

The impact of crashes. Crashes can occur anywhere and at any time, resulting in a range of outcomes, from serious injuries or fatalities to property damage only. When multiple crashes cluster at a location, that site is designated as a blackspot.

[1]   Royal Society for the Prevention of Accidents (1983) Crash Investigation Workshop notes, Cardington, England

B.   Breaking the chain of events leading to a crash

13.   There are many varied types of road crashes. Some involve two or more motor vehicles, some involve one motor vehicle, while some involve pedestrians, or cyclists. Invariably, a road crash is the culmination of a chain of events. A chain is made up of links; by looking broadly at the links within the chain of events, a blackspot team can determine the most effective way to break one or more of them. This means that similar crashes in the future can be prevented.

"Blackspot teams aim to break a link in the chain of events that leads to the crashes"

14.   By considering a hypothetical crash and examining the chain of events leading to it, one can explore the logic required by a blackspot team (Box 1).

15.   It is easy to attribute this crash to alcohol, driver inattention, or speeding. But the new brakes performed differently from what was expected, and may have been a contributing factor. The snowy conditions might have impaired visibility, and icy roads are common in such cold weather. But more importantly, if the bridge works had been adequately signposted well in advance, the driver might have had sufficient time to safely respond to the changing road conditions ahead. Moreover, had a safety barrier been installed to shield the old bridge railing and the side slope, the car might not have ended up in the river, significantly reducing the crash's severity.

16.   Road safety engineers will say that each factor played a role in the crash, emphasizing that removing just one of these factors (breaking one link in the chain of events) could have averted this hypothetical crash.

17.   They approach a crash logically, focusing on the links within their sphere of influence to break. While they may not be able to directly influence a fatigued young male driver's state of mind or vehicle design and maintenance standards, they can critically assess the road conditions, the absence of warning signs, the use of unforgiving items to block a traffic lane, and the exposed parapet (a roadside hazard). By doing so, they can recommend a package of quick and cost-effective countermeasures to break at least one link, likely preventing future crashes like this.

"Blackspot investigations are reactive — they aim to reduce the number of future crashes at a site"

18.   Whether investigating a single serious crash (such as this hypothetical crash) or examining a blackspot with multiple crashes, an experienced team of road safety engineers will focus on contributory factors within the road and the environment.

19.   They will seek treatments that are effective in reducing the likelihood of a crash. These treatments may include sound road safety engineering inputs in highway designs, enhanced delineation, improved signs, clearer traffic control, and lower speed limits.

20.   They will also consider treatments that can influence exposure to a crash, such as redirecting traffic away from the site (reducing exposure) or banning certain turns at intersections (eliminating exposure to certain turning-related crashes).

21.   Finally, they will try to identify treatments known to mitigate the severity of crash outcomes. Depending on the type of crashes, these measures may include installing suitable safety barriers, removing roadside hazards, reducing speed limits, using humped crossings, or constructing roundabouts. Each of these treatments conforms to the broad category of acceptable treatments under the Safe System.

22.   The Safe System now serves as the guiding road safety philosophy for major road organizations and international development agencies. It is an important philosophy for everyone involved in road safety, as outlined in Chapter 1, Section C. Experienced blackspot teams will look for treatments that comply with Safe System principles to ensure they obtain maximum safety benefits from their work.

Box 1: The Chain of Events Leading to a Hypothetical Crash

It's a Monday morning, and a young man drops off his car at the mechanic for a service. They agree he should return to pick it up after work at 5 p.m. He has had a tiring weekend and is going through a rough patch with his boss at work.

He arrives at the garage at 5 p.m. only to discover his car has brake problems, and the repair will be more costly and time-consuming than initially thought. He texts his wife, explaining that he will be home late. She replies that she will visit her sister and be home at 9 p.m., and he should find his own meal.

Just after 6 p.m., he gets into his car and decides to grab a quick meal at a nearby restaurant. Upon entering the restaurant, he is surprised to meet old friends from his football days. They invite him to join them, and their socializing lasts longer than planned. Alcohol is consumed, but no one pays much attention to the quantity.

Upon leaving the restaurant, they find it is snowing outside. The young man offers his friends a ride as they don't have a car. They relax during the drive, reminiscing about their days as local football champions. Only the driver wears a seat belt.

Traveling around a sweeping right-hand curve on a road he has not traveled in months, he is taken by surprise by bridge works. There are no advance warning signs, only two small signs at the bridge. One lane on the bridge is closed with concrete blocks and a hazard board. He brakes, but the new brakes, replaced just hours earlier, grab and the rear of his vehicle slides sideways. It strikes the bridge railing and rolls down the steep side slope into the small river below.

Emergency services are only alerted when another vehicle passes by the scene over an hour after the crash. The two passengers in the rear seat have been ejected; one has passed away, and the other sustains serious injuries, requiring months of hospitalization. The front seat passenger is also seriously injured, while the driver suffers only minor injuries.

Source: Asian Development Bank road safety engineering consultant.

C. The Safe System approach to road safety

23. The Safe System approach to road safety incorporates the following principles (Figure 1):[2]

- Death and serious injuries are unacceptable. There should be no deaths or serious injuries on the roads. The Safe System approach prioritizes the elimination of crashes resulting in death and serious injuries.
- Humans make mistakes. People will inevitably make mistakes and decisions that may contribute to crashes, but the road transport system can be designed and operated to accommodate human mistakes and to avoid death and serious injuries when a crash occurs.
- Humans are vulnerable. The human body can only withstand limited forces during a collision before death or serious injuries occur. Therefore, designing and operating a road system that accommodates physical human vulnerabilities is crucial.
- Safety is a shared responsibility. All stakeholders, including road authorities, traffic police, government agencies, industry, and the public, play a significant role in preventing fatalities and serious injuries.
- Safety is proactive. Proactive tools should be used to identify and address safety issues in the road transport system, rather than waiting for crashes to happen.
- Redundancy is crucial. Reducing risk requires strengthening all components of the road transport system so that if one part fails, the other parts will still protect people.

"The Safe System aims to create a transport system that is more human proof and incorporates infrastructure that considers human vulnerabilities to prevent fatal or serious injury crashes"

Figure 1: Diagrammatic Representation of the Safe System

Source: Federal Highway Administration. 2022. Making our Roads Safer through a Safe System Approach. *Public Roads - Winter 2022*. 85 (4).

24. The Safe System is an approach to road safety that requires roads to be managed to prevent death and serious injuries. It developed in the early years of this century, amalgamating proven aspects of earlier road safety philosophies.

25. The Safe System recognizes the fallibility of humans, who can only withstand limited kinetic forces during a crash before incurring serious or fatal injuries. It also acknowledges that humans will inevitably make mistakes on the roads. The United Nations, the World Health Organization, and major development banks have endorsed the Safe System as the foundational philosophy for the Second Decade of Action in Road Safety 2021–2030.

26. The goal of the Safe System is to establish a transport system that is more resilient to human factors. The Safe System expects road infrastructure to be designed and maintained with consideration of human vulnerabilities to minimize the risk of fatal

---

[2] Federal Highway Administration. 2022. Making our Roads Safer through a Safe System Approach. *Public Roads - Winter 2022*. 85 (4).

or serious injury crashes. Furthermore, it expects a shared responsibility for averting fatal and serious crash outcomes, with road authorities playing a significant role in achieving these objectives. It is considered unacceptable to blame road users for crashes when infrastructure improvements can reduce these risks.

D.   Applying the Safe System

27.   While it has generally been a priority for most highway authorities to focus on programs addressing severe crash outcomes, the Safe System places an even greater emphasis on addressing fatal and serious crash outcomes.

28.   Crash risk is the product of three elements: likelihood, exposure, and severity.[3] Those involved in a national blackspot removal program will recognize the importance of considering these three elements as they investigate blackspots and seek to apply Safe System compliant treatments. When assessing each element at a known crash site, the team will determine which of these elements to focus on at the site. Some general road safety options related to these three elements, some of which fall under the responsibility of professionals other than road safety engineers, include the following:

29.   Influencing the *likelihood* of a crash by:

- implementing evidence-based road safety engineering techniques/devices in treatments for high-frequency crash sites and new road designs,
- conducting road safety audits of new road designs,
- performing road safety inspections of the existing road network,
- establishing and enforcing appropriate speed limits,
- modifying road user behavior through road design and public awareness,
- launching effective public awareness campaigns, and
- consistently enforcing road rules.

30.   Influencing the *exposure* to a crash by:

- promoting safer forms of transport, such as cars over motorcycles; and
- constructing separate footpaths and bicycle paths/lanes.

31.   Influencing the *severity* of a crash by:

- ensuring consistent, safe speed management;
- designing intersections compliant with the Safe System;
- creating forgiving roadside environments, with safety barriers where necessary;
- providing high-quality post-trauma care; and
- promoting seat belt wearing in motor vehicles, and helmet use by motorcyclists.

> "Blackspot teams aim to identify options that reduce the likelihood of future crashes, exposure to future crashes, and the severity of any such crashes that may occur"

32.   To prevent serious road trauma, the Safe System requires the forces during collisions to remain below the limits that the human body can tolerate. To minimize these forces during an impact, the Safe System focuses on the following:

- **Speed management.** If casualty crashes are to be prevented, speeds must be managed first to minimize the risk of a collision and second to reduce impact forces to levels survivable by humans.
- **Intersection design.** The Safe System emphasizes the safe design of intersections to minimize relative impact angles and impact speeds, reducing the impact forces on vehicle occupants.
- **Forgiving roadsides.** Providing forgiving roadsides is crucial to minimize the consequences when a vehicle leaves the road.

33.   Safe speeds are integral to the Safe System approach; they influence crash causation and significantly affect crash severity. Research has shown a clear relationship between the outcomes for a given speed environment and the survival impact speeds for different crash types.

---

[3]   Austroads. 2021. *Guide to Road Safety Part 2: Safe Roads.*

34.   Figure 2 demonstrates that at impact speeds above 30 kilometers per hour (km/h), the chance of survival following collisions between vehicles and pedestrians reduces dramatically.

35.   Figure 2 also illustrates that the survival impact speed for side impacts at intersections is 50 km/h, and for head-on crashes, it is 70 km/h. When considering serious injury risk, these speed thresholds decrease. For example, the equivalent speed for serious injury is only 20 km/h for pedestrian collisions, 30 km/h for side impacts (near side), and also 30 km/h for a head-on collision.

36.   These research findings indicate that for the elimination of death and serious injury in road crashes, one of the following measures is necessary: (i) providing road infrastructure to prevent such crash types from occurring (likelihood and/or exposure), (ii) improving vehicle crashworthiness (likelihood and/or severity), or (iii) reducing impact speeds to these Safe System thresholds (severity). As an example, this entails installing median separation to prevent head-on crashes on two-way highways, or reducing operating speeds to 70 km/h or lower where there is no median separation.

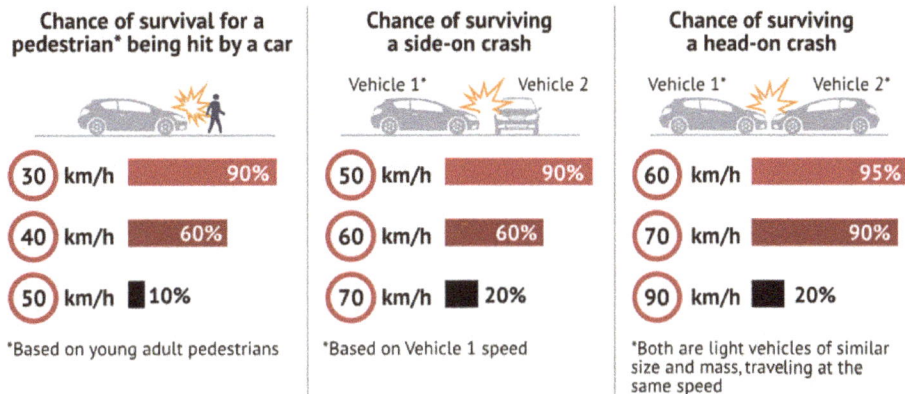

Figure 2: Chance of Survival in Three Crash Types at Three Impact Speeds

Chance of survival for a pedestrian* being hit by a car	Chance of surviving a side-on crash	Chance of surviving a head-on crash
30 km/h — 90%	50 km/h — 90%	60 km/h — 95%
40 km/h — 60%	60 km/h — 60%	70 km/h — 90%
50 km/h — 10%	70 km/h — 20%	90 km/h — 20%
*Based on young adult pedestrians	*Based on Vehicle 1 speed	*Both are light vehicles of similar size and mass, traveling at the same speed

Source: Based on Wramborg, P. 2005. "A new approach to a safe and sustainable road structure and street design for urban areas." Road safety on four continents conference. 2005. Warsaw, Poland. Swedish National Road and Transport Research Institute (VTI). Linkoeping, Sweden.

Crash risk on two-way roads. As traffic volumes increase on a two-lane, two-way road, the risk of a head-on crash also increases, prompting the Safe System to seek measures to prevent head-on crashes on these roads, such as separating traffic flows or reducing speeds to 70 kilometers per hour or lower in the absence of median separation.

Reducing the risk of head-on crashes. Wire rope safety barrier has been installed to separate opposing traffic streams and prevent head-on crashes.

"A vital aspect of the Safe System is to manage impact speeds to minimize the risk of fatal and serious casualty crashes"

37.   The Haddon Matrix can help put this into perspective (Table 1). It is a commonly used paradigm in the field of injury prevention. Developed by William Haddon, this matrix examines factors related to personal attributes, mechanical attributes, and environmental attributes before, during, and after an incident resulting in death or injury. By considering this matrix, an investigator can evaluate the relative importance of different factors and interventions.

38.   When analyzing road crashes, the Haddon Matrix considers three main factors (vehicle, human, road) and three phases (precrash, crash, postcrash). Understanding these phases allows the development of suitable crash countermeasures.

39.   For example, road safety engineering countermeasures can be devised to decrease the likelihood of a crash (precrash phase) and subsequently reduce crash severity (crash phase) in the event of a collision. After that, road safety engineering may, to a lesser extent, assist access for emergency vehicles in the postcrash phase.

40.   For the hypothetical crash previously described, the following factors can be applied across its three phases in the Haddon Matrix (Table 2).

41.   For detailed information on safe speed management, please refer to CAREC Road Safety Engineering Manual 7 (Speed Management).

Table 1: Haddon Matrix Example

Phase	Human Factors	Vehicles and Equipment Factors	Environmental Factors
Precrash	Information Attitudes/awareness Impairment Police enforcement	Roadworthiness Lights, indicators Electronic stability control Brakes including ABS	Road design and road layout Speed management Delineation Pedestrian facilities Intersection controls
Crash	Use of seat belts Impairment	Occupant restraints Safety devices (air bags) Crash-protective design	Protective roadside devices Forgiving roadsides
Postcrash	First-aid skills Access to medics Impairment	Ease of access Fire risk	Rescue facilities Traffic congestion

ABS = antilock braking system.
Source: Wikipedia. Haddon Matrix (last modified 19 October 2023). https://en.wikipedia.org/wiki/Haddon_Matrix.

Table 2: A Haddon Matrix for a Hypothetical Crash

Factors	Precrash	During Crash	Postcrash
Vehicle		New brakes	
Human	Fatigue Possible alcohol consumption Potentially excessive speed	Failure to wear seat belts	Delay in alerting postcrash trauma care
Road	Possible presence of ice Absence of advance road work signs Hazardous concrete blocks on the bridge Exposed bridge parapet	Possible presence of ice Lack of suitable safety barriers to prevent vehicle from falling into the river	Vehicle not easily visible from the road

Source: Asian Development Bank road safety engineering consultant.

42.   The historical crash data of a particular site has long been acknowledged as a reliable predictor of future road crashes at that site. This has been the underpinning rationale for blackspot removal programs in numerous countries spanning many years. By analyzing the crash history of a site, an experienced blackspot team can recommend changes to the road and its environment that will reduce the likelihood and/or severity of future crashes at the site.

"Blackspot removal programs are about crash reduction"

43.   The blackspot process encompasses three key stages: identification, investigation, and treatment.

44.   Figure 3 outlines the blackspot process, showing the stages of identification, investigation, and treatment.

A.   Identification

45.   The responsibility of collecting crash data falls upon the traffic police. The amount of data gathered for a crash may vary, but it usually includes time and location data (crash-specific data), person data (about the people involved), vehicle data (about the vehicles involved), and road data (about the location). Location data are particularly important for blackspot investigators, and the widespread use of mobile phones, many equipped with GPS capabilities, allows police to pinpoint the crash site accurately. Other critical data, such as the directions of the vehicles or road users and the prevailing conditions at the time of the crash, are usually obtained from witnesses and/or crash victims.

46.   Depending on the system in place in each country, the crash report form is either maintained as a paper copy (with a duplicate sent to police headquarters) or converted into a digital format and stored in a national crash database. Typically, a national crash database is managed and maintained by the traffic police, the Ministry of Internal Affairs, or a similar agency.

47.   For an effective blackspot removal program, or for the effective treatment of a single high-crash frequency site, sharing crash data with key stakeholders, including the national road authority and other relevant parties, is essential. Global experience has shown clear advantages in sharing crash data among these stakeholders. The data shared need not include personal or private details, such as names, ages, addresses, or license information. Such personal details are necessary for the traffic police for prosecution purposes, but are irrelevant to road authority engineers and other stakeholders who are primarily concerned with site and vehicle-related factors.

"Knowing the exact location of a blackspot requires accurate crash data"

48.   The shared data are interrogated at agreed intervals, usually every 6 or 12 months, by the stakeholder responsible for the national blackspot program. While this stakeholder is likely to be the national road authority, it may also be the traffic police. From the interrogated database, sites across the country with the highest number of reported casualty crashes are listed, and these sites are considered for possible inclusion in the ongoing national blackspot removal program.

49.   The list should have more sites than can be funded, as some sites may not proceed to treatment. This may be due to several reasons:

- The number of crashes may be reducing year by year, and it may be prudent to monitor the site for another year to observe if this trend persists.
- The site may have been previously investigated, with no effective action possible despite a consistent history of crashes.
- Some sites may lack a viable pattern of crashes.
- The location may have been changed under another program (such as being bypassed by a new road), reducing traffic usage and crash exposure.

Figure 3: The Blackspot Process

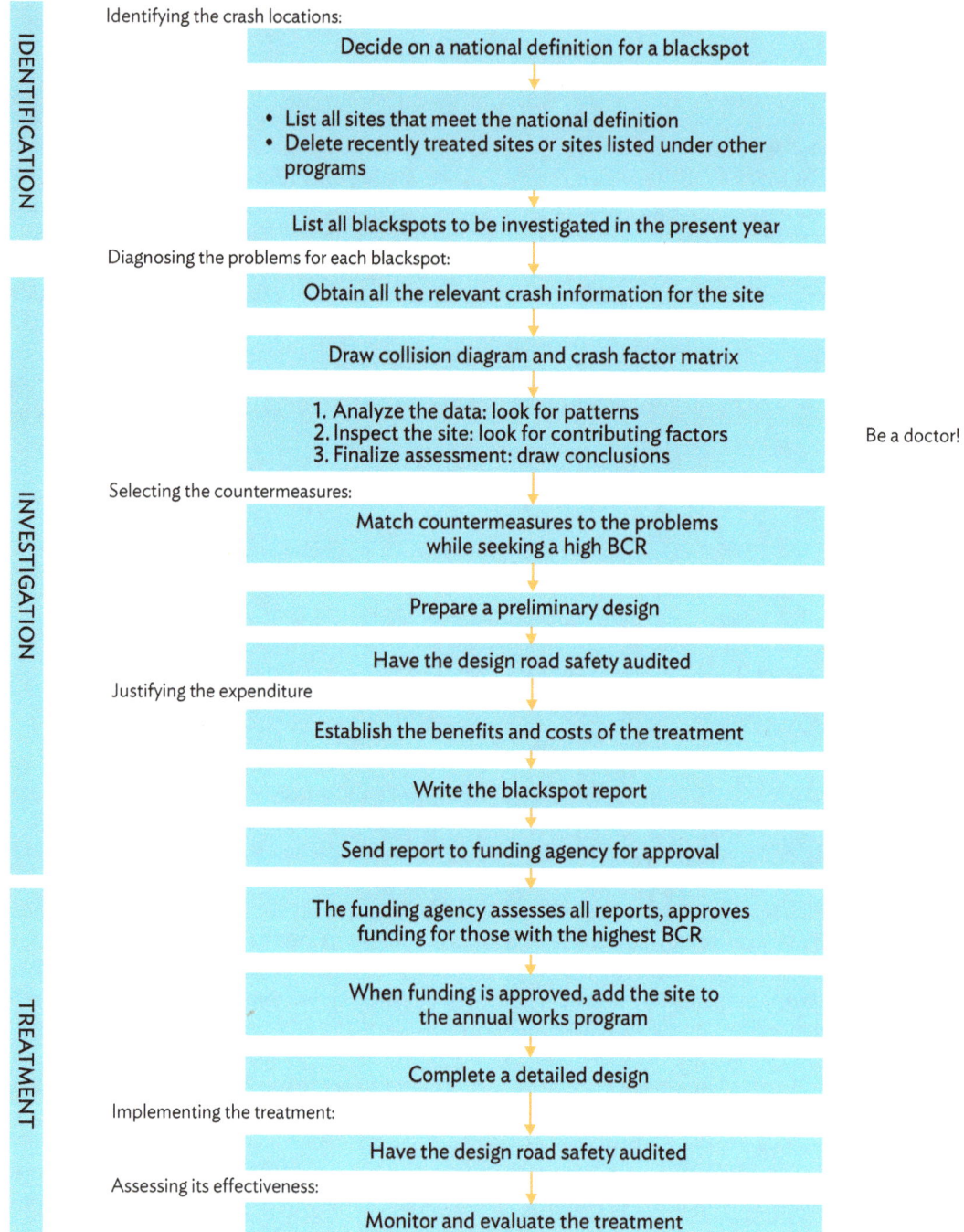

IDENTIFICATION

Identifying the crash locations:

Decide on a national definition for a blackspot

- List all sites that meet the national definition
- Delete recently treated sites or sites listed under other programs

List all blackspots to be investigated in the present year

Diagnosing the problems for each blackspot:

Obtain all the relevant crash information for the site

Draw collision diagram and crash factor matrix

1. Analyze the data: look for patterns
2. Inspect the site: look for contributing factors
3. Finalize assessment: draw conclusions

Be a doctor!

INVESTIGATION

Selecting the countermeasures:

Match countermeasures to the problems while seeking a high BCR

Prepare a preliminary design

Have the design road safety audited

Justifying the expenditure

Establish the benefits and costs of the treatment

Write the blackspot report

Send report to funding agency for approval

The funding agency assesses all reports, approves funding for those with the highest BCR

TREATMENT

When funding is approved, add the site to the annual works program

Complete a detailed design

Implementing the treatment:

Have the design road safety audited

Assessing its effectiveness:

Monitor and evaluate the treatment

BCR = benefit–cost ratio.
Source: AUSTROADS. 2021. *Guide to Road Safety Part 2: Safe Roads AGRS02-21*. Sydney (modified by the Asian Development Bank road safety engineering consultant).

- Resettlement or environmental issues may hinder remedial actions.
- Utilities and services become apparent during design and slow down progress.

50. Preparation then begins for the sites on the agreed national list of blackspots to be investigated in the second stage of the blackspot process.

B. Investigation

51. The investigation stage is pivotal in the blackspot process, demanding substantial investigative skills from the blackspot team. It comprises six key steps:

Diagnosis

52. Starting with the sites at the top of the national list, the blackspot team collects comprehensive crash data for each site. If the national crash database is well developed and generates detailed information on each crash (including the exact location, severity, vehicle/pedestrian direction, time of day, day of the week, weather, and light conditions), the blackspot team can promptly diagnose the issues at each site. However, if the national database is not as well developed and lacks this level of detail, the blackspot team should request copies of individual crash report forms from the traffic police to extract the necessary data. Only by having sufficient, accurate crash data can the team create a collision diagram and a crash factor matrix (see Chapter 4) for each site. The collision diagram and crash factor matrix are essential tools for the diagnosis of any blackspot. They help the blackspot team identify crash patterns at a site.

Site inspection

53. Following the preparation and examination of the collision diagram and crash factor matrix, the blackspot team inspects the site. The main purpose of the site inspection is to closely examine the road and traffic environment for factors contributing to the crash patterns. Nighttime inspections are highly recommended, especially if many reported crashes occur at night.

List of possible countermeasures

54. A list of potential countermeasures is then developed based on the contributing factors identified at the site and the blackspot team's experience with similar crash patterns. The selection of countermeasures considers the crash pattern(s) identified in the diagnosis phase and prioritizes "primary" Safe System treatments, which are proven to greatly reduce the risk of death or serious injury in specific crash scenarios, over "supporting" treatments, which mostly address crash likelihood. Selecting suitable countermeasures can be time-consuming, often requiring numerous cost–benefit iterations before the team agrees on the optimal set of countermeasures.

> "The site inspection looks for contributing factors to the crashes, and these can guide the team toward appropriate countermeasures"

Concept design

55. Once a package of countermeasures has been agreed upon by the blackspot team, a concept design is prepared to ensure the package's practicality and to estimate its cost. The concept design should undergo a road safety audit to provide an independent safety check.

Economic assessment

56. The next step in the investigation stage involves an economic assessment of the recommended package of countermeasures. The cost and estimated benefits of the treatments are calculated to determine a benefit–cost ratio (BCR) for the package. The BCR is used by the funding agency to prioritize funding for blackspots in the annual program.

Blackspot report preparation

57.   The various documents, reports, collision diagrams, and crash factor matrices prepared during the investigation stage are compiled into a blackspot report. This report should be formatted to facilitate the assessment of the site and its package of countermeasures against other blackspots in the national blackspot removal program. It should also present how the BCR for the recommended package of countermeasures was calculated.

58.   The funding agency, usually at the national road authority's head office, uses the BCRs in the blackspot reports to determine the sites to be funded first and the sites which must wait for future funding.

C.   Treatment

59.   Once a site secures funding approval, the concept design progresses to detailed design. During this phase, the work is costed with greater precision, and the drawings undergo another round of road safety audit.

60.   Most blackspot countermeasures are installed/constructed while the road remains open to traffic. This places a heavy responsibility on both the contractor and the road authority to prioritize safety during the road works. Manual 2, "Safer Road Works," in the CAREC road safety engineering series offers guidance on this important aspect.

61.   Upon completion of the works, where feasible, a preopening road safety audit should be conducted. Manual 1, "Road Safety Audit," in the CAREC road safety engineering series offers guidance on the proactive process of road safety audit.

Monitoring

62.   Monitoring the performance of road safety treatments post-implementation is an essential yet often overlooked activity. Annual monitoring of crash performance can ascertain the positive and negative effects of the treatment. The new data can also contribute to before/after evaluations of the treatment as part of a wider treatment program to develop crash reduction factors (CRFs) detailed further in Chapter 6, Section C. Over time, this process improves the accuracy and confidence in predicting a treatment's effectiveness at new blackspots.

63.   Monitoring of the site should continue for several years from the date of implementation. If traffic police crash data continue to reveal casualty crashes at the site, the blackspot team should explore all available options.

64.   In due course, as experience grows within a national blackspot removal program and more sites are treated, the research team can analyze before-and-after crash statistics, potentially leading to the development of a national set of CRFs. The same research team may also evaluate the benefits of the national blackspot removal program. International experience consistently demonstrates $4 in benefits in crash savings to the community for every $1 spent at the site.[4]

---

[4]   Corben, B., Newstead, S., Diamantopoulou, K. & Cameron, M. 1996. *Results of an evaluation of TAC funded accident black spot treatment.* Combined 18th Australian Road Research Board Transport Research Conference and Transit New Zealand Land Transport Symposium. 18 (5). pp. 343–359.

III.   Identifying Blackspots

A.   What is a blackspot?

65.   The prevalence of crashes in specific areas and particular crash types at single locations typically indicate common underlying causes. A blackspot removal program aims to identify and investigate these common causes, and subsequently apply appropriate countermeasures to reduce such crashes in the future. In essence, a blackspot removal program is focused on crash reduction.

66.   A blackspot is a location on the road network with a high number of casualty crashes. The locations can take various forms (Figure 4):

- An individual site (e.g., an intersection or a curve)
- A length of road (urban or rural)
- An area of the road network (such as a local traffic area)
- Locations throughout the road network that share common hazardous features (e.g., Y-junctions) and/or common crash types (e.g., incidents involving young pedestrians). These locations are usually addressed through mass action treatments (see Chapter 5, Section E).

"A blackspot is a location on the road network with a high number of casualty crashes"

67.   The criteria for defining a blackspot can vary significantly from one country to another. Most countries with a national blackspot removal program require a certain number of recorded casualty crashes within a specified time frame. For example, a country might define a location as a blackspot if it experiences 12 or more casualty crashes in the most recent 3-year reporting period.

68.   However, the necessity of the program and budget considerations can lead a country to adjust these figures. A country with a good road safety record might set a relatively low threshold for the number of casualty crashes to compile a sizeable list of candidate blackspot sites. It simply lacks locations with a high incidence of crashes. Conversely, countries with poorer road safety performances may have numerous sites with a high number of casualty crashes. To generate a manageable list of blackspots, they may set a relatively high benchmark for the number of crashes required to qualify as a blackspot.

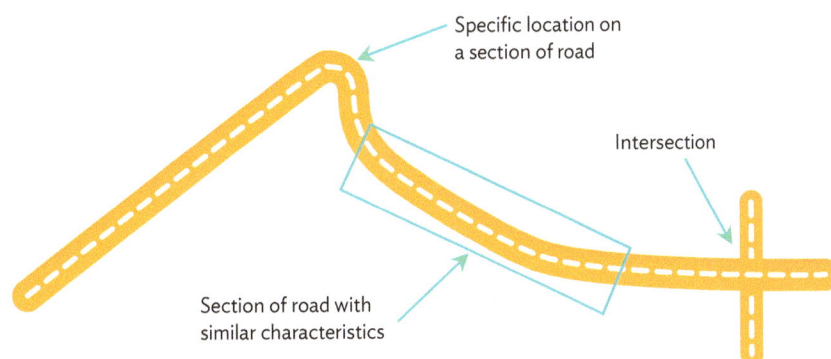

Figure 4: Diagrammatic Representation of Possible Blackspots

Specific location on a section of road

Intersection

Section of road with similar characteristics

Source: Asian Development Bank road safety engineering consultant.

"For road authorities in CAREC countries, it is recommended that blackspots be defined by a number of casualty crashes within a 3-year period"

69.   Some countries employ a scoring system for each crash at a site, with scores based on crash severity. For example, a fatal crash might receive 10 points, a serious casualty crash 5 points, and a casualty or property crash 1 point (Table 3). Adding up these points for all the crashes at a location yields a total, which is then compared with other sites across the country. Starting from the top and working down the list, this method sets a benchmark when there are sufficient sites to be investigated to fully utilize the annual blackspot budget. A positive aspect of this method is that it directs more attention to locations where fatal and serious casualty crashes have occurred, aligning with the Safe System approach.

70.   In some cases, a smaller number of road authorities have attempted to define their blackspots according to the number of casualty crashes per the number of vehicles using the site. This crash rate definition often favors low-volume intersections (a few hundred vehicles per day), where a small number of casualty crashes can result in a high crash rate relative to busier sites with more crashes. This method can lead to blackspot funds being spent on low-volume sites at the expense of busier sites, which may raise questions from road users. Additionally, having fewer crashes to investigate makes it more challenging to identify crash patterns, which is why the use of this definition has largely been discontinued.

71.   A few countries define a blackspot as any location where a fatal crash has occurred in the past 3 years. This approach is often followed by countries with less robust crash data systems, as fatal crashes tend to be better reported than other types. But this definition is not recommended as it allocates significant resources to individual crashes, many of which may be the only crash at that site. Consequently, there is little or no opportunity to detect crash patterns at these sites, and recommendations are likely based on limited data.

72.   There is also a strong argument that some minor collisions become fatal crashes simply because seat belts (or crash helmets for motorcyclists) were not worn at the time. While seat belts and helmets are essential for road safety, blackspot investigations aim to identify parts of the road and roadside environment that have failed the road users. This reinforces the importance of an effective national blackspot removal program focusing on sites with a high number of reported casualty crashes.

73.   The definition of a blackspot is a prerogative of the national road authority, and it is always preferable to identify a blackspot as a location on a road with a large number of casualty crashes.

74.   For road authorities in CAREC countries, an agreed-upon number of casualty crashes within a 3-year period is recommended. This is a straightforward and easily understandable definition. While the benchmark figure of casualty crashes and the time frame may vary among countries, this method has been tried and tested and can serve as a valuable starting point for CAREC road authorities in initiating a national blackspot removal program.

Table 3: Blackspots Ranked by the Crash Severity Scoring System

Site	Fatal Crashes (10 points)	Serious Casualty Crashes (5 points)	Other Crashes (1 point)	Total Points	Priority Ranking
A	2	7	4	59	3
B	0	18	2	92	1
C	0	10	4	54	4
D	0	7	6	41	5
E	3	8	2	72	2

Source: Asian Development Bank road safety engineering consultant.

75.   In the future, road authorities may need to adjust the threshold for the number of casualty crashes needed to qualify a site for a national blackspot removal program. Typically, as a successful blackspot removal program reduces the number of sites exceeding the benchmark (in the pool of blackspots), the necessary qualifying number of crashes decreases.

76.   As an example, an Australian state launched a blackspot removal program in 1980, initially defining a blackspot as any site with at least 12 casualty crashes in the previous 3 years. Over time, this state has lowered the qualifying threshold for a blackspot to just three casualty crashes in a 5-year period. This definition is just 15% of the first qualifying benchmark, signifying success in a program that has seen hundreds of blackspots treated during that time. However, it also implies that most of the remaining sites, with such low numbers of casualty crashes, lack discernable crash patterns to guide the blackspot team in developing cost-effective countermeasures.

77.   There is a universal desire to promptly address the worst blackspots in a country, creating pressure to prioritize the identification of these sites. Regardless of the chosen system, it should be noted that identifying blackspots is just the first step in the blackspot investigation process.

B.   Collaboration between traffic police and engineers

78.   There is a fundamental difference between the types of crash investigations carried out by traffic police and road safety engineers. Traffic police investigate serious crashes individually, scrutinizing details such as brake marks, impact speeds, and minute indications of vehicle failure. Road rule breaches are also a large part of these detailed investigations to identify and prosecute the offender.

79.   In contrast, engineers assess multiple crashes that have occurred at a specific site over the previous (generally) 3 years. They analyze patterns within these crashes in search of one or possibly two dominant crash patterns that could benefit from treatment with proven countermeasures.

80.   Blackspot removal programs rely on accurate police crash data and the dedicated investigative work of blackspot engineers. The highest levels of success are achieved when traffic police and engineers collaborate closely within a coordinated blackspot removal program. Such collaboration is enhanced when these two professions have a clear understanding of each other's roles and openly share information.

Defining blackspots in Central Asia Regional Economic Cooperation. For the Central Asia Regional Economic Cooperation road, a recommended definition of a blackspot is a location with at least 12 casualty crashes within the past 3 years.

81.   On some occasions, traffic police join engineers during a site inspection. These joint inspections have many benefits, especially if the officers have attended crashes at the blackspot, providing valuable learning opportunities for both groups. On other occasions, traffic police and engineers may convene to discuss individual sites or the overall progress of the national blackspot removal program.

82.   Both groups have significant and sometimes challenging roles in road safety. It is beneficial for them to understand the challenges faced by the other. For instance, when a crash occurs and traffic police respond to the scene, they have several crucial tasks. Their most urgent responsibilities include aiding victims and ensuring the site's safety for others. This may involve traffic control, first aid, and potentially lifesaving measures if an ambulance has not yet arrived. They might also need to prevent passersby from smoking if there is spilled petrol near the crashed vehicle(s). Only when these most pressing tasks are under control can the traffic police officers begin gathering crash data and interviewing witnesses (if any).

"Traffic police need personal details of those involved in a crash for prosecuting offenders, but other stakeholders (like engineers) do not"

83.   Among these often-demanding tasks, the one of utmost importance for a blackspot program is gathering crash data. Many traffic police collect crash data with the primary aim of using it for legal prosecution. However, it is essential for them to realize that the crash data they gather also play a vital role in guiding engineers toward treatments that can effectively reduce the number and severity of future crashes. Reliable and accurate crash data are a cornerstone of an effective blackspot program. Indeed, the crash reduction work undertaken by blackspot teams can lighten the future workload of traffic police.

84.   Without accurate crash data, it becomes difficult for engineers to identify blackspots, and subsequently, it becomes more difficult to discern crash patterns in the site's data. This, in turn, hampers the development of viable countermeasures to reduce the crash problem.

85.   Blackspot investigations are best carried out by a small team of professionals, usually engineers from the national road authority or contracted by it. A small team provides shared knowledge and multiple pairs of eyes during inspections, enhancing the potential to detect contributing factors and leading to the best possible package of crash countermeasures. Collaborative work in a team also presents valuable mentoring opportunities for expanding road safety engineering knowledge and skills. Encouraging the development of these skills in CAREC is essential, as most CAREC road authorities face a shortage of road safety engineers, with few local consultants specializing in this field.

86.   Members of a blackspot team normally possess the following:

- Training in road safety engineering, covering topics such as the blackspot investigation process, cost-effective countermeasures, roadside hazard management, signs and markings, pedestrian facilities, safe geometric design, and more.
- Knowledge and skills in traffic engineering and/or road design.
- Knowledge of the Safe System approach to road safety.
- Empathy for the safety of all road users, regardless of age or gender.
- Sound judgement, including the ability to apply logic to complex, dynamic situations.
- Effective management and communication skills to collaborate within the blackspot team, liaise with traffic police and other departments in the road authority, and coordinate the preparation of the blackspot report.

87.   While traffic police and engineers have distinct roles in the blackspot process, they share a common link in regularly sharing accurate crash data. This data exchange is a critical activity for every blackspot removal program. Currently, it represents a weakness in many CAREC countries. Enhancing the collection and storage of crash data and its subsequent sharing among stakeholders (excluding personal details of those people involved) is a task under review in some countries, requiring greater efforts.

C. The need for good crash data

88.   Ensuring accuracy and clarity is paramount when describing various aspects of crashes, such as fatalities, fatal crashes, casualties, and casualty crashes. One fatal crash can involve multiple fatalities and may also encompass casualties alongside the fatality or fatalities. Crashes are classified based on their maximum casualty level, spanning from fatal crashes to serious casualty crashes, casualty crashes, and finally property damage-only crashes (Table 4).

89.   These definitions for crash severities are consistent with those used in most countries and by most major development agencies. It's worth noting that some countries may have slightly different definitions; for example, some may define a fatality resulting from a crash as occurring within 7 days from the time of the incident.

"Good crash data, shared among stakeholders, are the essential foundation for road safety activities in any country"

D. What if crash data are insufficient?

90.   A successful blackspot removal program needs access to reliable, accurate, and complete crash data. While traffic police may have good crash data for high-frequency crash locations, there are instances when data are limited or entirely absent.

91.   Blackspot teams and other users of road crash data should be aware of data limitations and, if possible, seek to resolve any anomalies that may arise. Common limitations include unreported crashes and inaccuracies in crash report forms and the coding system. It is widely acknowledged that more severe crashes (fatal and serious casualty crashes) are more likely to be reported to police than crashes involving lesser or no injuries. According to a New Zealand study, only 60% of serious casualty crashes made it into the crash database, with the percentage significantly lower for minor injury crashes.[5] Nevertheless, there are other ways in which crash data may be lacking:

- Crashes involving illegal activity, such as underage driving, driving while intoxicated, or driving without a license, often go unreported.
- Reporting errors can occur because the police often have more pressing needs at a crash scene than filling out a crash report form. Delays in form completion can introduce inaccuracies. Additionally, reporting officers may lack adequate training in crash reporting, leading to incorrect data entries. For example, if a crash form requires identifying a possible contributing cause for the crash, an inexperienced police officer may cite "failing to give way," whereas a more experienced officer (or a road safety engineer) might identify the cause as "control sign not visible."
- Coding errors can occur when the paper copy of the crash report is coded into the database. Common errors arise when the person responsible for coding uses an incorrect definition for classifying crash number or assigns the wrong direction for one of the road users (traveling east instead of west).

Table 4: Crash Types

Crash Severity		Description
Casualty crashes	Fatal	A crash resulting in the death of one (or more) people within 30 days of the crash
	Serious casualty	A crash that does not result in a fatality but involves at least one person with an injury serious enough to require hospital admission
	Slight casualty	A crash that does not result in a fatality or serious injury, but in which at least one person sustains slight injuries not requiring hospital admission
	Damage only	A crash in which there are no casualties; only vehicles or property is damaged

Source: Asian Development Bank road safety engineering consultant.

---

[5]   Alsop, P. and Longley, D. 2001. *Under-Reporting of Motor Vehicle Traffic Crash Victims in New Zealand. Accident Analysis and Prevention.* 33 (3).

- Location errors occur if the location information on the crash report form is imprecise or if the road has two names. Listing a crash under one name may lead to a location error or potentially underestimate the actual number of crashes at a location.

"Limited crash data hinder good, effective blackspot removal programs, as well as their future growth and development"

92.   When a country lacks complete, accurate crash data, or only has limited information available, or records crash data in a broad manner (for instance, documenting crashes by kilometer distance along a road rather than at specific points), it becomes a challenge for a blackspot team to:

- identify locations with the highest crash frequencies across the country,
- pinpoint areas with the most casualty crashes,
- gather detailed information about the crashes at specific sites,
- develop patterns of crashes with the limited data at hand, and
- make informed decisions about cost-effective countermeasures for reducing casualty crashes at a site.

93.   Consequently, it becomes uncertain whether the worst blackspots in the country are being identified and investigated first or whether the most cost-effective treatments are being considered to address the actual crash patterns at these blackspots.

94.   In addition, without a baseline number of crashes, calculating the reduction in crashes due to the countermeasures implemented at a site becomes unfeasible. This can lead to subsequent debates about the effectiveness of these treatments, leading to varying opinions on what works well and what does not.

95.   Consequently, a long-term drawback emerges: if the crash reductions generated by selected countermeasures cannot be accurately calculated, it will not be possible to develop a set of CRFs for

the country. CRFs are the percentage reduction in casualty crashes due to a particular countermeasure. A full table of CRFs is provided in Appendix 1 of this manual, with more details on the significance of CRFs given in Chapter 6.

"Treating and eliminating a blackspot is an important task, and every opportunity should be taken to gather as much crash data as possible to facilitate a thorough and accurate investigation"

96.   Having crash data for casualty crashes at the site for the most recent 3 years is a standard requirement for an effective blackspot investigation. International experience with blackspot programs has shown that multiple years of accurate crash data are usually necessary for identifying crash patterns. However, in many CAREC countries today, complete and detailed crash data are not readily available. In such situations, what should a blackspot team do?

97.   When investigating high-frequency crash sites or any problematic location where complete crash data are lacking, it is essential to explore alternative options. Treating and eliminating a blackspot is an important task, and every opportunity should be taken to gather as much crash data as possible to facilitate a thorough and accurate investigation.

98.   Therefore, if crash data for the last 3 years (or the period specified in the national blackspot definition) are unavailable, efforts should be made to obtain the most recent 2 years of crash data. If this is not feasible, even a single year of accurate crash data can prove useful in a blackspot investigation.

99.   In cases where only a limited amount of crash data is available, it may be possible to add data from other sources such as hospitals and local health-care centers; coroners' reports; or discussions with local government officers, nearby residents, farmers, or business proprietors. Additional sources may include crash reports maintained by road managers in their

Box 2: A Story of Perseverance by a Blackspot Team

In a European city, a large roundabout on an elevated road comes to the attention of the blackspot team. It has a history of nighttime crashes, typically occurring between 12 midnight and 2 a.m., with vehicles approaching from the north colliding with the central island. Although it is a clear pattern of crashes, only a few result in casualties.

The blackspot team inspects the site during the day and finds little to report. Back in the office, they discuss the site and review the photographs they have taken. They conclude that they need to inspect the site during the actual hours of the crashes, which are mostly late at night. Given that it is winter and darkness falls around 4:30 p.m., the team decides to inspect the site during the evening hours but ensures they leave early to be home for dinner.

The evening inspection yields no substantial findings.

The next day, the team once again discusses the site. Some members contemplate redesigning the roundabout, while others suggest moving to another site since this one does not have many serious casualties. Some even consider that the problems might be related to the drivers, possibly intoxicated, as the city's hotels and restaurants close at midnight. But the team leader recognizes that they have not completed one of the most critical tasks—to inspect the site during the actual crash times. The site exhibits a clear late-night pattern, and they resolve to wrap up warmly and inspect the site between midnight and 2 a.m. the following night.

Sitting in a layby just before midnight, the team still finds nothing noteworthy. However, as the clock strikes midnight, half of the streetlights suddenly turn off. Driving from the north, the team observes that the remaining streetlights create the illusion of a straight road, with the central island of the roundabout concealed in the shadows. Even attentive drivers might not have sufficient time to spot the island and avoid a collision.

A quick call to the power supply company the next morning reveals that the streetlights are turned off every night at midnight as part of an energy conservation program. A prompt agreement between the supply company and the road authority ensures that a selected number of lights near the roundabout remain illuminated throughout the night. This action quickly and effectively resolves the crash problem.

The low-cost removal of this blackspot would not have happened without the perseverance of the blackspot team and their experienced team leader.

Source: Asian Development Bank road safety engineering consultant.

maintenance records or information from media reports (newspapers and social media). While this type of data gathering may be time-consuming and potentially less accurate, it can guide the blackspot team toward understanding the crash problems at the site, ultimately leading to practical treatment options.

100. If no reported crash data are available but the site has a reputation as a high-frequency crash location, a final option is to subject it to a road safety inspection (commonly referred to as a road safety audit of an existing road). An experienced road safety inspection team can reveal deficiencies at the location and propose remedial improvements based on their expertise and judgement. The road safety inspection team should focus on aspects of the road and the roadside that may be contributing to suspected crashes. They should look for missing components such as delineation, warning signs, or line markings, as well as other items needed to warn, guide, control, or assist road users at the site. Box 2 illustrates a blackspot team's determination to solve a recurring nighttime crash problem.

101.   The road safety inspection team should rely on their experience to ensure that the issues they report directly relate to the suspected crashes. The team should avoid merely upgrading the location with standard devices unless those devices are recognized as valid crash countermeasures.

102.   Box 3 explains the concept of "restart" and "overshoot" issues in right-angle collisions at crossroad intersections and offers effective treatments for addressing each issue.

Box 3: Restart vs. Overshoot

Right-angle collisions at crossroad intersections may be attributed to either a restart issue or an overshoot issue.

Restart occurs when the driver is aware of the intersection, slows down, perhaps stops, but then proceeds forward into an "unsafe" gap, resulting in a right-angle crash.

The driver knows the intersection is there but makes a wrong decision moving forward. Why? Are sightlines obstructed? If so, is this due to vegetation, parked vehicles, or structures like bus shelters, with either vertical or horizontal geometry?

The objective of addressing this type of collision is to improve sightlines, which assists with correct decision-making. Typical effective treatments for addressing restart collisions include:

- improving safe intersection sight distance,
- maximizing sightlines,
- reducing operating speeds on the main road,
- implementing geometric changes to improve sight distance (such as reducing crests),
- trimming trees and grass to improve sight distance,
- reversing the traffic control at the intersection (although this may introduce other risks), and
- introducing a roundabout or traffic signals.

Overshoot occurs when the driver does not know the intersection is there and passes through it at normal operating speed. When this happens, it leads to a right-angle collision if another vehicle is simultaneously passing through the intersection on the intersecting road.

Why would a driver overshoot an intersection? It's possible that the driver is not paying attention or is distracted. But it may be due to limited sight distance to the intersection caused by road geometry, vegetation, or parking. It may also be due to see-through issues, where a driver's view is drawn to an object in the distance, diverting attention from the immediate intersection.

If the blackspot team suspects an overshoot problem at the blackspot, one way to address this type of collision is to make the intersection more conspicuous. Effective treatments for addressing overshoot collisions include:

- improving approach sight distance,
- installing advance warning signs and/or advance direction signs,
- duplicating the stop or give way signs,
- using pavement markings and/or colored pavement,
- adding lighting (only if crashes are at night), and
- installing a roundabout or traffic signals.

Source: Asian Development Bank road safety engineering consultant.

"Investigating blackspots requires time, skill, and determination. During this stage, the blackspot team overlays the crash history on the blackspot site and seeks effective countermeasures"

103. After establishing a national list of blackspots or when an individual site comes under scrutiny due to a recent serious crash, the investigation stage of the blackspot process begins. This stage can be both interesting and frustrating, depending on the site's characteristics, the accuracy of crash data, and other relevant factors.

104. This stage can also be a rewarding part of the blackspot process, when the team uncovers crash patterns at a site and can propose practical treatment options to mitigate future incidents. It is an important stage where the investigating engineers need to apply their skills, carefully examining crash patterns and site conditions to identify contributing factors that can lead to effective countermeasures.

A. The steps in the investigation stage

105. The investigation stage of the blackspot process includes the following steps:

- **Diagnosis.** This step involves sourcing detailed crash information from individual crash report forms and finding crash patterns via collision diagrams and crash factor matrices.
- **Site inspection.** During this phase, investigators look for contributing environmental factors at the crash site and, if possible, at the times when the main crash patterns occur.

- **List of potential countermeasures.** The team aligns potential countermeasures with Safe System objectives to reduce the severity of dominant crash patterns and/or the likelihood of further crashes.
- **Concept design.** This step entails producing a concept design for the recommended package of treatments.
- **Economic assessment.** An economic assessment of the package of treatments is carried out to support funding requests within the national blackspot removal program.
- **Blackspot report.** The final step is preparing a comprehensive blackspot report.

106. Figure 5 illustrates the key steps in the blackspot investigation stage.

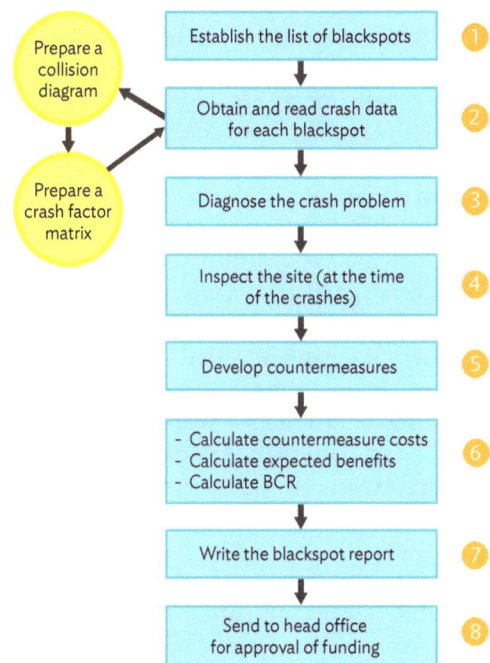

Figure 5: The Eight Key Steps in the Investigation Stage of the Blackspot Process

Prepare a collision diagram

Prepare a crash factor matrix

1. Establish the list of blackspots
2. Obtain and read crash data for each blackspot
3. Diagnose the crash problem
4. Inspect the site (at the time of the crashes)
5. Develop countermeasures
6. - Calculate countermeasure costs
   - Calculate expected benefits
   - Calculate BCR
7. Write the blackspot report
8. Send to head office for approval of funding

BCR = benefit–cost ratio.
Source: Asian Development Bank road safety engineering consultant.

B.   Diagnosis

107.   Starting with the sites at the top of the national list, the blackspot team collects as much information as possible about the crashes at each site. If the national crash database is well developed and generates full details for each individual crash—such as precise location, severity, vehicle or pedestrian direction, time of day, day of week, weather, and light conditions—the blackspot team can readily access the data for diagnosing site-specific issues and for preparing a collision diagram and crash factor matrix.

108.   However, if the national database is less developed and lacks detailed crash data, the blackspot team should request copies of individual crash report forms from the traffic police. This will assist the extraction of essential data required for diagnosis.

109.   The blackspot team should thoroughly examine each report form, with a keen eye for recurring details, such as crashes involving particular types of vehicles (e.g., buses, trucks, motorcycles) or those originating from specific directions (e.g., approaching downhill from the north). While it is rare to detect clear crash patterns this way, this approach aids in forming a broad overview of the site's crash history.

110.   A collision diagram and a crash factor matrix (discussed in the next section) serve as indispensable investigative tools for diagnosing a blackspot, helping the blackspot team identify recurring crash patterns at a site.

Investigative tools

111.   Two tools are available to assist the team in uncovering crash patterns at a site. The first, a collision diagram, illustrates the directions of travel for vehicles and road users at the moment of the crash, generating crash patterns based on these movements. It is a very useful tool, but it has certain limitations, such as its inability to depict patterns of nighttime crashes or incidents occurring in adverse weather conditions like rain or on icy roads.

112.   Here is where the crash factor matrix plays an important role. It highlights patterns derived from crash data items not represented by the collision diagram. These patterns include light conditions (day/night/dusk), road conditions (wet/dry/icy), and time of day or day of week. These tools provide the

blackspot team with visual records of crash details, aiding in the identification of patterns.

113.   A third investigative tool, a crash chart, reveals trends in crashes over time. These charts prove valuable for displaying aggregate crash information and are particularly useful when analyzing severity trends over time. For example, they can effectively display the results of monitoring crashes before and after improving a site or throughout an annual blackspot program.

114.   Details of these three valuable investigative tools are provided below.

Collision diagrams

115.   A collision diagram is a visual representation of the vehicles and other road users involved in crashes, depicting the direction in which each was traveling at the time of the crash. This diagram is created using information retrieved from the road traffic crash database and/or individual police crash report forms.

116.   Two collision diagrams are shown below. Figure 6 displays a clear pattern of crashes at a crossroad intersection, while Figure 7 presents a less obvious pattern of collisions at a median opening. Both diagrams are valuable, but the second one may require more time to discern the underlying patterns.

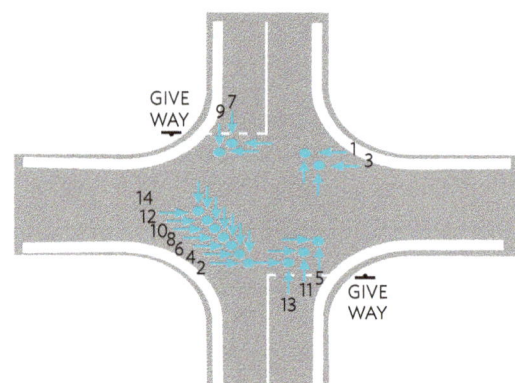

Figure 6: A Collision Diagram for a Blackspot at a Crossroad Intersection

Note: This collision diagram illustrates a clear pattern of right-angle collisions, with 9 out of 14 crashes involving vehicles from the north.
Source: Asian Development Bank road safety engineering consultant.

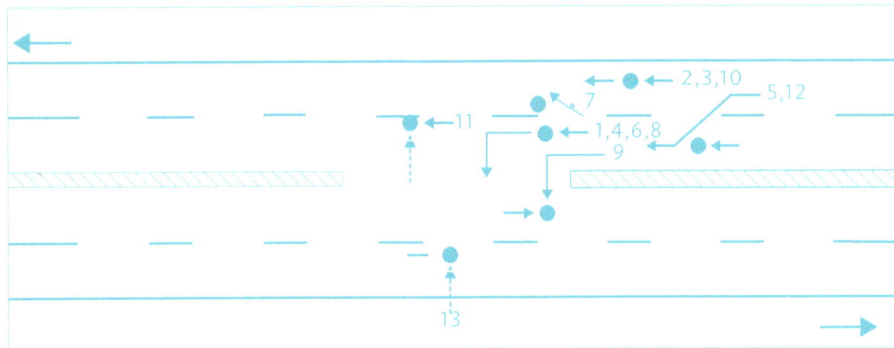

Figure 7: A Collision Diagram for a Blackspot at a Median Opening on a Busy Divided Highway

Note: There is a cluster of crashes involving through traffic and vehicles that are slowing down to make a U-turn at the median opening.
Source: Asian Development Bank road safety engineering consultant.

117.   Collision diagrams can be prepared manually or using a computer application. They:

- depict each crash through arrows and/or symbols representing the involved road users,
- indicate the directions of these road users at the time of impact,
- rely on accurate crash data, as an error in the crash report, such as misdirection (north instead of south), can hinder the collision diagram's ability to reveal meaningful patterns, and
- focus attention on groups of similar crash types that are grouped together.

118.   Additionally, collision diagrams may include crash severity. For instance, fatal and serious injury (FSI) crashes may be denoted by a dark circle and a light circle, respectively, at the collision point. Other crash details, including the date, time, weather conditions, and light conditions, can be added as labels to the arrows. In practice, however, it can be difficult to identify patterns in these textual descriptions. This difficulty leads to the development of the crash factor matrix.

Crash factor matrices

119.   A crash factor matrix is a table of summarized crash data, where each column corresponds to one crash. This format proves instrumental in blackspot analysis by unveiling potential crash patterns related to factors like severity, time, day of week, weather conditions, and light conditions, which are not apparent from a collision diagram (Table 5).

120.   Occasionally, these matrices are referred to as a crash factor grid, a crash factor table, or even a "stick diagram" due to the use of small "stick figures" to represent each crash (Tables 6 and 7). They are an invaluable tool for use in blackspot investigations.

Crash charts

121.   Crash data can be examined in alternative ways by visually plotting individual data points to observe changes over time (Figure 8). Crash charts serve as a third investigative tool that can help highlight particular crash features and types at a blackspot, including:

- crash trends and severity over time,
- crash types (according to the classification definitions, as discussed later),
- time of day or day of the week,
- types of vehicle involved,
- road surface conditions,
- weather conditions, and
- light conditions.

Table 5: Crash Factor Matrix for Blackspot in the Collision Diagram in Figure 6

Crash Number	1	2	3	4	5	6	7	8	9	10	11	12	13	14
Date: Month	3/06	04/10	19/11	08/06	03/07	07/11	30/12	27/02	03/05	24/07	18/04	21/05	14/06	20/08
Day of week	Sat	Wed	Thurs	Sun	Thurs	Fri	Tue	Fri	Sun	Fri	Sun	Fri	Mon	Fri
Time of day	1700	1855	1530	1900	1345	2145	1900	1220	1800	2000	1845	1610	1735	1855
Severity	3	3	2	3	2	4	3	3	4	2	3	2	2	3
Light conditions	gray	black	yellow	black	yellow	black	yellow	black	gray	black	yellow	yellow	gray	black
Road conditions	Wet	Wet	Dry	Dry	Dry	Dry	Dry	Dry	Dry	Dry	Dry	Dry	Wet	Dry
DCC code	110	110	110	110	110	110	110	110	110	110	110	110	110	110
Object 1	Car	Car	Car	Car	Car	Car	Car	Car	Car	Car	Car	Car	Van	Car
Object 2	Car	Car	Truck	Car	Car	Car	Car	Truck	Car	Car	Car	Car	Car	Car
Object 3					Car			Car			Car			
Direction 1	N	S	N	S	N	S	S	S	S	S	N	S	N	S
Direction 2 (& 3)	W	E	W	E	E	E	W	E	W	E	E	E	E	E
Other														

DCC = description for classifying crashes.

Note: Night = black; day = yellow; dusk/dawn = gray.

Source: Asian Development Bank road safety engineering consultant.

Table 6: Crash Factor Matrix for a Site with Crashes in Chronological Order

Crash Number	1	2	3	4	5	6	7	8	9	10	11	12	13	14	15
Day of week	Mon	Tue	Wed	Sat	Mon	Sat	Wed	Mon	Sun	Fri	Mon	Tue	Tue	Thu	Sat
Date: Month	12/06	18/06	04/07	05/10	18/02	12/03	15/11	06/12	18/03	22/04	10/05	25/05	15/07	18/09	12/11
Time	16.10	9.10	18.15	10.15	16.40	19.10	11.21	18.45	13.45	15.10	8.15	9.11	19.20	12.00	16.10
Light conditions	yellow	black	yellow	yellow	yellow	black	black	black	yellow	yellow	yellow	yellow	black	yellow	yellow
Road surface	Wet	Wet	Dry	Wet	Dry	Wet	Wet	Dry	Wet	Dry	Dry	Wet	Dry	Dry	Dry
Severity	2	3	2	3	3	3	1	3	3	1	2	3	2	2	2
DCC code	110	130	122	172	130	130	110	173	110	102	122	160	100	110	110

DCC = description for classifying crashes.

Note: Patterns do not stand out readily in this example. Table 7 (below) shows crash patterns for this site more readily.

Night = black; day = yellow; dusk/dawn = gray

Source: Asian Development Bank road safety engineering consultant.

122.   The most useful application of a crash chart is to illustrate the changing numbers of casualty crashes at a site over time. If the number of casualty crashes at a blackspot appears to decrease each year, even if the reason is unknown, it may be reasonable to set that blackspot aside from the current program and allocate resources to other sites with ongoing or increasing crash numbers. A site temporarily removed from the program should not be forgotten but should be subject to regular monitoring. If the crash frequency begins to rise, it should be reinstated on the blackspot list for investigation.

Table 7: Crash Patterns Based on Description for Classifying Crashes Codes

Crash Number	7	1	14	9	15	3	11	2	6	5
Day of week	Wed	Mon	Thu	Sun	Sat	Wed	Mon	Tue	Sat	Mon
Date: Month	15/11	12/06	18/09	18/03	12/11	04/07	10/05	18/06	12/03	18/02
Time	11.21	16.10	12.00	13.45	16.10	18.15	8.15	9.10	19.10	16.40
Light conditions										
Road surface	Wet	Wet	Dry	Wet	Dry	Dry	Dry	Wet	Wet	Dry
Severity	1	2	2	3	2	2	2	3	3	3
DCC code	110	110	110	110	110	122	122	130	130	130

DCC = description for classifying crashes.

Note: Night = black; day = yellow; dusk/dawn = gray.

Source: Asian Development Bank road safety engineering consultant.

Figure 8: Crash Chart Example—Trends in Crash Frequency and Severity

Year	Fatal	Serious Injury	Slight Injury	Total
1	1	2	1	4
2		2	3	5
3	1	3	3	7
Total	2	7	7	16
Mean crash rate per year				5.3

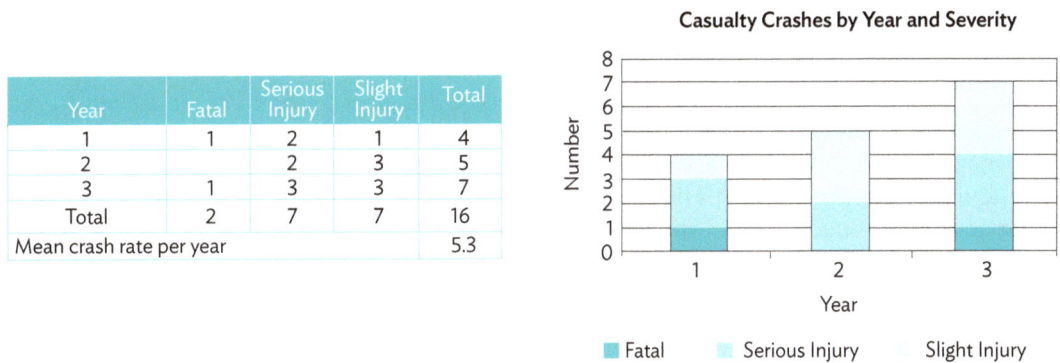

Casualty Crashes by Year and Severity

Source: Asian Development Bank road safety engineering consultant.

123.   Experience has shown that most crash patterns can be uncovered using a collision diagram in conjunction with a crash factor matrix. For this reason, these are the two main tools used by blackspot teams during the investigation stage. Crash charts are useful for showing crash trends over time at a site or group of sites and can be especially beneficial for monitoring the effectiveness of a blackspot program.

124.   Box 4 highlights the analogy between blackspot engineers and doctors, emphasizing the importance of accurate diagnosis and effective remedies in addressing road safety issues.

Box 4: Comparing Blackspot Engineers to Doctors: A Diagnostic Approach

At times, drawing parallels between the work of a blackspot engineer and that of a doctor can be insightful.

Doctors must accurately diagnose a patient's illness before prescribing the correct medicine and providing treatment to facilitate their recovery. In some cases, this diagnostic stage can be a matter of life and death. No one wants a doctor to make a misdiagnosis, as prescribing the wrong medicine is wasteful at best and harmful at worst.

A blackspot can be likened to a sick part of the road network; it requires a remedy to restore its health. But the blackspot investigation team cannot ask questions of the blackspot; they must rely on their observational skills to identify contributing factors at the scene. Subsequently, they must ensure that the package of countermeasures they prescribe will be effective.

Source: Asian Development Bank road safety engineering consultant.

C.   Site inspection

125.   Once the collision diagram and the crash factor matrix are prepared, and the crash data have been diagnosed as thoroughly as possible, it is time for the blackspot team to inspect the site. Whether the location is nearby or on the other side of the country, the site must be inspected. Nighttime inspections are highly desirable, and if the reported crashes occur mostly at night, then a nighttime inspection is essential.

"Whether the location is nearby or on the other side of the country, the site must be inspected"

126.   The main purpose of the site inspection is to identify contributing factors to the crash patterns in the road and traffic environment. Team members should put themselves in the shoes of the crash-involved road users and try to understand what those people may have been doing, thinking, or seeing at the moment of their crash.

127.   At this point in a blackspot investigation, engineers need to apply their broad experience and knowledge of crashes. They must look for features in the road environment that might mislead or confuse some road users. They need to ask why some people had collisions at the site, but the majority did not. They need to empathize with the crash-involved persons and consider how the collision could have occurred.

128.   The inspection of a blackspot is an essential part of the blackspot investigation process. The objectives of a site inspection are to:

- identify site characteristics that may contribute to crashes at the site (further guidance is provided below),
- observe traffic operation and road user behavior at the site,
- relate the results of the diagnosis (crash patterns) to road user behavior and road features at the site, and
- begin to consider suitable countermeasure options for reducing the frequency of crashes and/or reducing the severity of the crashes at the site.

"The blackspot team visits the site not to witness crashes but to examine the environment, monitor traffic, and identify factors that could contribute to crashes"

129.   Ideally, the site inspection should take place close to the times of the main crash patterns. If there are nighttime patterns, the main inspection should take place at night. If the crash pattern typically involves weekends, the site inspection should include a weekend visit. While the physical road environment doesn't change from hour to hour, traffic volumes, speeds, and behaviors may vary, and light conditions certainly do. Aligning the site inspection with the general time periods for the dominant crash pattern(s) is a useful approach. While it is highly unlikely that the blackspot team will witness a crash, this approach affords them the opportunity to monitor traffic behavior during critical times and possibly observe other routine activities, such as markets (e.g., local farmer's markets or street vendors), schools (student drop-offs or pickups), and public transportation movements (e.g., bus stops).

130.   As a blackspot team often has multiple sites to investigate, it is essential for team members to keep the following points in mind during their inspection schedule:

- Allocate ample time on-site. Observing the behavior of all road users is time-intensive, with site inspections usually taking hours. Multiple visits may be necessary.
- Review the collision diagram and crash factor matrix on-site. These tools can guide the team toward potential contributing factors to the crash problem.
- Consider contributing factors related to intersection layout, road design, roadside features, sight distances, sign visibility, traffic speed, traffic control (e.g., give way or stop signs, traffic signals, roundabouts), and road surface conditions.
- Address all groups of road users. Gain a pedestrian perspective by walking around the site and crossing the road, especially if pedestrians are part of the crash patterns.

- Extend the inspection beyond the immediate vicinity of the blackspot. This is to ensure that contributing factors approaching the site can be identified, which is crucial for blackspots on rural highway curves or blacklengths.
- Engage in discussions with local residents or workers near the blackspot to gather their insights and ideas. Local knowledge can assist in reaching conclusions about factors contributing to crashes.
- Use any available site plans to record observations and site measurements/features, as well as to sketch treatment options.
- Take photographs and videos of the site from all approaches. These visuals are useful for demonstrating existing safety concerns and for further review in the office. High-quality images support recommendations in the blackspot report. Photographs and videos can also demonstrate before-and-after scenarios, which are extremely useful for monitoring annual blackspot removal programs or securing funding for future program cycles. Decision-makers are always interested in successful examples of how the program budget is being used.

"While on-site, engineers should put themselves into the shoes of those involved in the crashes and try to understand their issues"

131.   The team should bear in mind that a single type of crash can have multiple causes. For example, a pattern of right-angle collisions at a give-way-controlled crossroad intersection may be due to the driver on the minor road either:

- not seeing the intersection and proceeding as if it was not there or
- knowing the intersection exists, stopping, but then making an error in gap selection.

132.   Both scenarios are plausible, and the site-specific road features and traffic volumes may influence the likelihood of one over the other. It is necessary to decide which option is more probable at the blackspot under inspection, as the treatment

Site inspections take time. During the site inspection, the blackspot team looks for items in the road environment that might mislead drivers or other road users.

for the two options can vary. Installing more advance warning signs for an intersection will not address the issue of poor gap selection, and relocating a bus stop on the major road to improve sight distance will not resolve an overshoot problem at the intersection.

133.　Making a list of suspected contributing factors at the site is a necessary part of the inspection process. Along with the photographs and videos taken on-site, this list serves as a useful record of the inspection findings. This list will later get used to develop the package of countermeasures, which will be costed, assessed, and included in the blackspot report. The blackspot report, along with the BCR of the recommended countermeasures, will determine whether funding approval is granted for the countermeasures. This is a fundamental step in every blackspot investigation. Without funding approval, blackspot treatments will not be implemented, and the crash problem will persist.

134.　A final, yet crucial, point to consider is personal safety during a site inspection. Ensuring the safety of the blackspot team members while on-site is of utmost importance.

135.　Team members should adhere to all necessary safety precautions as mandated by their employer, such as wearing reflective personal protective equipment, parking vehicles safely, and only entering the road when traffic conditions are clear.

136.　When taking photographs or measurements, it is essential to pay attention to traffic movements and avoid turning away from approaching vehicles. In busy or high-speed environments, having a colleague act as a "spotter" to monitor the traffic and provide warnings is advisable.

137.　At blackspots in high-speed areas, temporary traffic management measures like warning signs and alerting traffic control devices add an extra layer of safety for the team. In extreme situations, such as a blacklength on an expressway, inspections may have to be conducted from a moving vehicle traveling at or near expressway speeds. This may require many runs through the site. Alternatively, it could involve expressway closures although this is usually only practical on low-volume expressways and during selected times.

138.　Additionally, the team must ensure that their presence on-site does not distract road users in ways that might elevate the risk of collisions. Importantly, they must park their vehicles:

- legally and away from intersections, crests, and sharp curves,
- in positions that do not obstruct visibility,
- where they do not force pedestrians onto the roadway, and
- preferably off the road within designated parking areas.

D. List of potential countermeasures

139. From the list of contributing factors identified during the site inspection, the blackspot team proceeds to evaluate potential countermeasures for each one. They compile a list of potential countermeasures based largely on the blackspot team's experience with similar crash patterns observed at other sites. These countermeasures focus on the crash patterns identified in the collision diagram and the crash factor matrix.

140. Selecting suitable countermeasures can be time-consuming, sometimes extending over many days. It may involve multiple iterations of cost–benefit analyses before arriving at an optimum arrangement of countermeasures endorsed by the team. The team should focus on countermeasures that are proven to reduce fatal and serious crashes. This aligns with one of the primary objectives of the Safe System approach, which aims to mitigate crash severity. Although treatments that reduce crash likelihood should be considered, particularly if they offer a lower cost, the team should prioritize treatments offering the best reduction in fatal and serious crashes per dollar cost.

141. While most blackspots can be effectively addressed with conventional devices and treatments, a few may require innovative solutions. It is always desirable to maintain an open mindset when considering the package of recommended treatments. The team should raise broader questions such as the following:

- Are the countermeasures thoroughly tested and proven?
- Can the cost of the work fit within the budget of the national blackspot removal program? If not, should the project be deferred, or can it be divided into a short-term, low-cost first stage followed by a higher-cost final stage?
- Will the community readily comprehend and accept the proposed countermeasures?
- Could noncompliance by road users become a potential issue?
- Will the responsible agency support the ongoing management of the countermeasures? For example, if street lighting is a recommended countermeasure, which agency will cover the ongoing electricity and maintenance costs?
- Will the proposed treatment align with other treatments implemented elsewhere?

E. Concept design

142. Once the team agrees on the countermeasures, a concept design is drawn up to ensure the practicality of the countermeasure package and estimate its cost. The concept design should then undergo a road safety audit, particularly if the works involve substantial changes to the site, such as constructing a new roundabout.

F. Economic assessment

143. The next step in the blackspot investigation process is to undertake an economic assessment of the recommended countermeasure package. To do this, the team must first calculate the treatment costs (estimated design and construction costs) and the treatment benefits (estimated reduction in crash costs). Only after these calculations can the team determine the BCR for the package. Full details of this important step are provided in Chapter 6.

144. Selecting low-cost countermeasures with high CRFs will lead to high BCRs. The higher the BCR, the better the chances of obtaining funding for the remedial works.

> "Funding for blackspot removal works is usually approved based on the benefit–cost ratio calculated by the team"

G. Blackspot report

145. The documentation prepared during the investigation stage of the blackspot process is gathered and compiled into a report. The report should be in a format that allows assessment of this site and its package of countermeasures against all other blackspot projects within the national program. It should outline the actions taken in the investigation, its conclusions, and, importantly, the calculated BCR for the recommended countermeasure package.

146.   The blackspot report serves two primary purposes. First, it is submitted to the funding agency, usually the national road authority, to seek funding approval. The road authority will use the BCRs from all the blackspot reports it receives to determine which sites will receive funding for the current year and which will have to wait for future funding. Second, the report provides a record of the investigation to facilitate future monitoring of crashes at the site.

V. Treating a Blackspot with Suitable Countermeasures

147.   After completing the site inspection, the blackspot team should dedicate time to review the list of contributing factors they identified during their visit. From that list, the team considers possible countermeasures to mitigate future crashes at the site. This can be challenging, as there are often several treatments to evaluate for each contributing factor at a site. Determining which treatment to recommend over another is a decision that improves with experience, much like a doctor choosing the best medicine from several options for a sick patient.

148.   The blackspot team will usually have to consider practical issues that include:

- selecting countermeasures that target crash patterns,
- focusing on countermeasures that best address fatal and serious injury (FSI) crashes (in line with the Safe System),
- choosing options suitable for the traffic conditions at the site,
- evaluating feasible countermeasures that can fit within constraints posed by the site's terrain, nearby developments, utilities, services, and more, and
- assessing the overall cost of the countermeasure package.

149.   Noting that every blackspot in a national blackspot removal program must compete for funds before any countermeasures can be installed, the team should initially prioritize low-cost countermeasures. These offer the best chance of achieving a high BCR and, in turn, the best chance of approval for national funding.

150.   Chapter 1, Section D, explains that crash risk is the product of three elements: likelihood, exposure, and severity. The blackspot team will consider each of these three elements as they investigate a blackspot and seek to apply Safe System-compliant treatments. When evaluating a known crash site, the team will discuss which one, or more, of these elements to focus on.

151.   The team is aware that other professions also play essential roles in road safety. Schools, the media, traffic police, and the hospital system are all vital contributors to a successful road safety program. Increasing public awareness of the importance of obeying road rules, wearing seat belts (both front and back), adhering to speed limits, and refraining from using a mobile phone while driving are crucial tasks. Consistent police enforcement of speeding, drink-driving and seat belt wearing offences is of utmost importance within any national road safety strategy. High-quality post-trauma care can make the difference between life and death.

152.   The blackspot team may deliberate on whether to request increased police enforcement of speed at a blackspot or launch a general awareness campaign focused on increasing driver compliance with road rules at pedestrian crossings. However, more often than not, to reduce future crashes at a blackspot, the team will concentrate on the road and its environment. By improving the road and its environment, blackspot engineers can both prevent future crashes from occurring (likelihood) and reduce the severity of any crashes that do occur (severity).

153.   While it is not possible in this manual to offer a complete list of effective countermeasures suitable for every blackspot, certain crash types are better addressed by a relatively small range of treatments. Blackspots don't have a one-size-fits-all solution. Below are some examples of the issues a blackspot team may consider for major crash types. These lists are general and require the logic and judgement that an experienced blackspot team can bring to the investigation.

154.   Referring to the CRF table (Appendix 1) can also be valuable at this stage, as it provides an extensive and useful list of treatment options.

A.   Run-off-road crashes (typically in rural areas)

155.   If a blacklength has a pattern of run-off-road (ROR) crashes, a blackspot team may identify the following contributing factors during their route inspection:

- excessively high operating speeds,
- inadequate cross-sectional width, including a lack of paved shoulders,
- severe drop-offs from the traffic lane to the unpaved shoulder,
- inadequate superelevation and/or skid resistance, and
- poor or nonexistent delineation.

156.   Taking these factors into account, an experienced blackspot team will consider countermeasures aimed at reducing the likelihood of future crashes, decreasing the exposure to crashes, and mitigating the severity of any future crashes. These countermeasures may include the following:

157.   Influencing the *likelihood* of further ROR crashes by:

- implementing geometric changes to superelevation and horizontal alignment,
- widening the road cross-section,
- paving shoulders,
- improving pavement skid resistance,
- increasing sight distance to/through curves,
- lowering the speed limit, and
- enhancing delineation, including tactile edge lines and centerlines, reflective guideposts, chevron alignment markers, and warning signs.

158.   Influencing the *exposure* to a ROR crash by:

- constructing a new section of road on a new alignment to eliminate the hazardous blacklength (though this is rarely cost-effective).

159.   Influencing the *severity* of a ROR crash by:

- removing fixed roadside hazards,
- creating a forgiving roadside environment with a wide run-out area,
- installing correctly designed safety barriers, and
- managing a safe speed environment.

160.   The blackspot team considers all factors, recognizing that, even with a wider road, paved shoulders, good delineation, and tactile edge lines, some drivers may still make mistakes on the road. Therefore, the roadside should be designed to be forgiving of their errors, and the severity of any remaining crashes should be minimized through the use of safety barriers and clear run-out areas.

B.   Head-on crashes (typically in rural areas)

161.   When investigating a blacklength with a history of head-on collisions, the team might discover the following contributing factors during their site inspection:

- excessively high operating speeds,
- a narrow cross-section and a lack of paved shoulders,
- undulating alignment, which creates dips and crests that can conceal vehicles,
- restricted overtaking sight distance,
- high traffic volumes leading to a lack of suitable, safe overtaking opportunities, and
- obscured or missing centerline markings.

Run-off-road crashes. There are many reasons for run-off-road crashes. Experienced blackspot teams seek cost-effective treatments to minimize the frequency and the severity of these crashes.

Two-way highways present a head-on crash risk. To minimize head-on crash risk, consistent delineation coupled with geometric changes and speed management can be effective until the time the highway can be duplicated.

162.   An experienced blackspot team will consider countermeasures aimed at reducing the likelihood of future head-on crashes, decreasing the exposure to such crashes, and finally mitigating the severity of any future crashes. These countermeasures may include the following:

163.   Influencing the *likelihood* of a head-on crash by:

- prohibiting overtaking with signs and markings,
- increasing sight distance and opening sightlines,
- widening the road and paving the shoulders,
- constructing more overtaking lanes along the route,
- lowering the speed limit and increasing police enforcement, and
- improving advance warning, direction, and regulatory signs.

164.   Influencing the *exposure* to a head-on crash by:

- duplicating the road with a median or central barrier (this can be expensive, but it is a Safe System option worth considering),
- constructing more overtaking lanes along the route, and
- creating a one-way road.

165.   Influencing the *severity* of a crash by:

- managing speeds below the Safe System speed threshold.

166.   The blackspot team considers all three contributing factors, recognizing that even with a wider road and overtaking lanes, some drivers may still make mistakes by attempting unsafe overtaking. The road should reduce their exposure to head-on crashes, with a median and centerline barrier proven to be effective in achieving this. Such measures can be costly, so if a median is not feasible, the severity of any remaining crashes should be minimized. Lower speed limits (within the Safe System limit of 70 km/h for such collisions) will be needed and should be introduced and enforced.

C.   Intersection collisions

167.   If a blackspot at an intersection exhibits a pattern of right-angle crashes, the contributing factors that a team might identify during their inspection include:

- restricted sight distances,
- excessively high approach speeds,
- a see-through effect on a minor approach,
- obscured regulatory signs, direction signs, or traffic signal displays,
- insufficient safe gaps due to high traffic volumes (if controlled by stop/give way signs),
- intersection lacking conspicuity (strong shadows on sunny days), and
- inadequate street lighting (if a nighttime crash pattern).

168.   An experienced blackspot team will ask whether the crashes may indicate an overshoot problem or a restart problem at the intersection (see Box 3). They will then explore countermeasures to reduce the likelihood

Many blackspots occur at intersections. It is essential for safety that intersections are conspicuous to all approaching drivers and that the form of traffic control is clear and well understood by all. The risk of restart collisions and overshoot collisions (see Box 3) should be examined at intersection blackspots.

of future right-angle crashes at the intersection, minimize the exposure to such crashes, and finally reduce the severity of any such future crashes. These countermeasures may include the following:

169.   Influencing the *likelihood* of a right-angle crash by:

- reducing the risk of overshoot collisions with improved advance warning, direction, and regulatory signs,
- reducing the risk of restart crashes by opening sightlines and removing obstructions,
- renewing all line markings,
- redesigning the intersection, possibly introducing a roundabout or traffic signals, and
- improving street lighting (for nighttime crash patterns).

170.   Influencing the *exposure* to a right-angle crash by:

- closing one or more approaches to the intersection,
- implementing right-in/right-out control (or left-in/left-out as necessary) to minimize conflict points, and
- constructing a roundabout.

171.   Influencing the *severity* of a right-angle crash by:

- designing a Safe System-compliant intersection, such as a roundabout and
- managing safe speeds.

172.   The blackspot team considers all three contributing factors, recognizing that even with an improved intersection, some drivers may still make mistakes. The intersection should be forgiving of their errors and thus the severity of any remaining crashes should be minimized. Lower speeds (within the Safe System limit of 50 km/h for such collisions) are needed and should be introduced and enforced.

D.   Pedestrian crashes

173.   If the blackspot (mid-block or intersection) shows a pattern of pedestrian crashes, the contributing factors that a blackspot team might find during their inspection may include:

- excessively high approach speeds,
- restricted sight distance to/from the pedestrian crash locations,
- parking issues,
- insufficient safe gaps due to high traffic volumes,
- high-speed two-way traffic on multiple lanes,
- complex traffic movements,
- a wide road, lacking refuge islands or a median,
- inadequate street lighting (if a nighttime crash pattern), and
- no pedestrian facilities (signals).

174.   An experienced blackspot team will consider countermeasures aimed at reducing the likelihood of future pedestrian crashes, decreasing pedestrian exposure to a crash, and finally mitigating the severity of any such future crashes. These countermeasures may include the following.

Pedestrian crashes. Blackspots with pedestrian crashes are common, particularly in urban areas. A blackspot team should put itself into the shoes of the pedestrians, walk the site and seek to understand what may have caused these crashes. The team should remember the wide variety of pedestrians that use the site.

175. Influencing the *likelihood* of a pedestrian crash by:

- increasing sightlines to pedestrian crossing locations,
- banning parking,
- introducing pedestrian separation (time/space) treatments,
- installing curb extensions and/or pedestrian refuges (enhancing visibility),
- improving pedestrian facilities (such as pedestrian-operated signals),
- upgrading street lighting (for nighttime patterns), and
- enhancing advance warning and regulatory signs.

176. Influencing the *exposure* to a crash by:

- constructing footpaths and
- installing pedestrian fencing.

177. Influencing the *severity* of a crash by:

- implementing traffic calming measures,
- lowering speeds to the Safe System limit of 30 km/h for pedestrians, and
- designing Safe System-compliant intersections (if applicable).

178. The blackspot team considers all three contributing factors, recognizing that even with an improved location, some pedestrians and/or drivers may still make mistakes. The road environment should be forgiving of their errors, and thus the severity of any remaining crashes should be minimized by managing lower speeds (within the Safe System limit of 30 km/h for such collisions). For detailed information about safe

pedestrian facilities, please refer to CAREC Road Safety Engineering Manual 4 (Pedestrian Safety).

E. Mass action programs and treatments

179. The focus of a mass action program is on implementing specific countermeasures at multiple high-risk sites, rather than treating a single site with a package of countermeasures. Mass action programs apply proven countermeasures (such as chevron alignment markers, guideposts, roundabouts, safety barriers, and reduced speed limits) at a number of sites on the road network where a particular type of crash is suited for that countermeasure.

180. Examples of mass action programs (with proven countermeasures) include:

- addressing rural single-vehicle ROR crashes (by implementing paved shoulders and tactile edge lines),
- managing crashes with bridge structures (by installing safety barriers and improved delineation),
- handling crashes with utility poles (by removing or relocating poles, enhancing delineation, and improving skid resistance),
- replacing Y-junctions with roundabouts or T-junctions, and
- reducing pedestrian crashes (by increasing signal clearance times; upgrading street lighting; and implementing curb extensions, curb ramps, pedestrian refuges, new crossings).

181. An alternative approach to implementing a mass action program involves categorizing crashes based on the road user involved. For example, a road authority may decide to address crash sites involving child pedestrians. These crashes can be quite varied, including scenarios like road crossings, walking alongside the road, commuting to and from school, or shopping. Consequently, a package of mass action treatments may start by focusing on schools near busy roads. A consistent package of treatments may be proposed, which could include the construction of sidewalks; parking restrictions outside the school; the installation of a fully marked, signed, and lit pedestrian crossing; along with refuge islands and curb extensions. The work may be complemented with a package of warning signs and a child road safety education program.

182. In mass action programs, large numbers of sites receive the same treatment, even if not all of them have experienced crashes. Care needs to be taken when undertaking an economic assessment of a mass action program, as the CRFs applicable to such sites may differ from those where clusters of similar crashes occur (they are often lower). On the other hand, there will be economies of scale when implementing similar treatments, which will reduce the cost per unit installed.

Effective delineation for road safety. A consistent delineation of a route, achieved through the use of chevron alignment markers, edge lines, and reflective guideposts, is a low-cost and effective approach to keeping vehicles on the road and reducing run-off-road crashes.

Enhancing road safety with tactile edge lines. Tactile edge lines, whether raised or depressed, are proven and low-cost devices to mitigate run-off-road crashes, even in places where snow clearing is a concern.

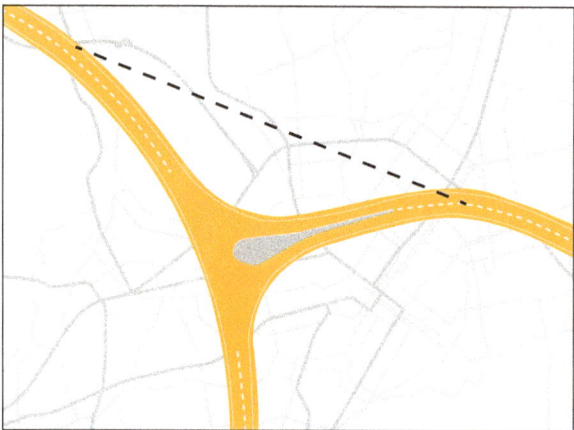

Y-junction removal or safer roads. The removal of Y-junctions, with a proven 85% crash reduction factor as seen in Appendix 1, will significantly reduce crashes and improve national road safety.

Enhancing bridge safety with simple measures. Reflective width markers, edge lines, and advance warning signs offer a cost-effective solution to reduce crashes at narrow bridges.

183. Mass action programs are particularly well-suited to CAREC countries, as they require less crash data to identify suitable sites, and mass action treatments can be implemented in large quantities to minimize costs. They are also ideal for multiple treatments along a route, enhancing driver confidence through the consistent application of mass action countermeasures.

184.   Road authorities bear the responsibility of wisely allocating government funds for the maximum good of the community. This accountability extends to funds allocated for blackspot removal. These funds must be used prudently to ensure that they deliver positive benefits to the community. When investing in the name of road safety, it is essential to assess whether the benefits will outweigh the costs involved.

"An economic assessment helps determine the most effective package of treatments for a single site"

185.   It is necessary to conduct an economic assessment for each package of proposed blackspot treatments. Several economic assessment models may be used, and the choice depends on the preference of the national road authority, with options such as the BCR, the net present worth model, and the first-year rate of return model.

186.   Each model has its own merits, along with varying degrees of accuracy and complexity. When using any of these models for an economic assessment of a proposed blackspot treatment, it is essential to keep in mind that the result is simply to assist the road authority in determining the priority for treating sites within a national blackspot removal program. An economic assessment serves as a method for guiding the prioritization of government expenditure in a blackspot removal program to maximize community returns.

187.   A commonly used model for blackspot assessments is the BCR. It is well understood by engineers and is straightforward to use once the benefits of a treatment have been calculated. It is commonly used by road authorities and recommended for use in blackspot removal programs. The BCR model for assessing blackspot treatments is detailed below.

A.   Economic assessment of blackspot treatments

188.   When investigating and treating an individual hazardous site, it is advisable to assess the economic costs and benefits of the proposed package of treatments to ensure that it delivers more benefits (in terms of crash savings) to the community than its cost. If the costs are projected to outweigh the estimated benefits, it is common to consider developing and assessing a lower-cost package of treatments, even if it may be less effective, or to contemplate leaving the site untreated and allocating funds elsewhere.

189.   When investigating multiple hazardous sites, such as within a national blackspot removal program, it is necessary to assess the costs and benefits of the recommended package of treatments for each of the blackspots to:

- ensure that the package yields more benefits (in terms of crash savings) to the community than it costs and
- assist the national road authority in prioritizing which blackspot to treat first, followed by the second, and so on, until the annual budget is fully used.

B.   Benefit–cost ratio model

190.   The BCR model is a widely used and straightforward economic assessment model applied not only in blackspot investigations but also in various fields. It calculates the ratio of the benefits (in $) to be gained by dividing it by the cost (in $) of doing it. If the ratio is 1.0 or higher, it signifies that the action is justified, as it yields more benefits than it costs.

191.   In practice, when managing a sizeable national blackspot budget, decision-makers often require a much higher BCR than 1.0 before approving funding. For new blackspot programs involving blackspots with many casualty crashes and relatively low-cost treatments, a higher BCR may be needed for funding approval. BCRs near or above 10 might be required

in some cases. On the other hand, a site in a well-established blackspot program will commonly obtain funding approval with a BCR higher than 4.

192. In a national blackspot removal program, comparing the BCRs from blackspot reports across the country allows the managing authority to prioritize which blackspot to fund first. Regardless of the total treatment cost, it is the BCR that determines the site's ranking for funding approval. This is because the BCR offers a relative measure of the benefits of the proposed treatment to the community.

"Comparing the benefit–cost ratios from blackspot reports across the country allows the managing authority to prioritize which blackspot to fund first"

193. Only when the site with the highest BCR involves very high costs, which in turn could consume most or all the annual blackspot program budget, do the managers of the national blackspot budget consider alternative options. These options might include staging the work over 2 or more years or deferring it until a larger budget becomes available. Another approach could involve treating a range of lower-cost treatments at numerous sites rather than allocating the entire budget to the first site. For instance, if the annual blackspot budget is $2,500,000 and the site with the highest BCR would consume the entire budget, the budget managers may choose to explore other options. They might decide to approve funding for several sites with lower BCRs but with much lower costs, collectively totaling $2,500,000. This approach would allow the budget to be spread across several districts. While the estimated number of crashes to be prevented may or may not be equivalent with this approach, it would display some uniformity in the provision of safer roads more widely across the nation.

194. Comparing the cost of a package of treatments with the predicted benefits of that package means it is necessary to express the treatment cost in the same units as the expected crash reduction benefits. Typically, this is done by calculating the cost (in

dollars or any currency) of the treatments, along with the value (in dollars or any currency) of the crash reductions the package of treatments is predicted to yield.

195. The BCR is calculated as follows:

$$BCR = \frac{\text{Benefits in \$ (casualty crashes prevented)}}{\text{Treatment cost in \$}}$$

196. The calculation requires various factors:

- the annual count of serious casualty crashes at the site,
- the CRF for the main selected crash countermeasure,
- the estimated life of the countermeasure (i.e., the number of years it is expected to provide safety benefits),
- the cost of the selected package of countermeasures, and
- the cost of a serious casualty crash.

197. Based on the given parameters, the formula is as follows:

$$BCR = \frac{\text{No. of crashes/year x CRF x project life x casualty crash cost}}{\text{Treatment cost in \$}}$$

198. In some cases, during the project development process, the ratios of various treatment options can be compared to determine the relative efficiency of the treatments in reducing serious crashes. While candidate projects with higher cost-effectiveness ratios generally provide greater road safety benefits, additional factors may also need to be considered as part of an iterative approach to the funding approval process.

Estimating the cost of the treatments

199. Estimating the cost of a package of treatments is a straightforward engineering task aligned with normal road authority practices. An initial cost estimate is formulated based on a concept plan and/or a description of works. It will generally include costs associated with:

- survey and design,
- construction, including traffic management,
- project management,
- the supply and installation of the treatments, and
- a contingency to account for uncertainties in the concept details.

Estimating the benefits of treatments

200. While calculating the cost of a package of treatments (in $) may be a routine engineering task, determining the likely benefits (also in $) of the package of treatments for a blackspot is a more abstract activity, particularly for most CAREC road authorities. This requires two main pieces of information to be available to blackspot investigators:

- the agreed cost (in $) for a casualty crash in the respective country and
- a CRF for the main countermeasure proposed for the blackspot.

201. Estimating the cost of a casualty crash (these include FSI crashes) can be sourced from various outlets. Development banks, universities, and selected crash research organizations have methodologies for calculating casualty crash costs.

202. If a country lacks an established cost for a casualty crash (which is the case for most CAREC countries), one way of costing a casualty crash is to adopt the method proposed by McMahon and Dahdah.[6] Their report states that a fatality in any country costs 70 times the gross domestic product (GDP) per capita of that country. It goes on to say that a serious casualty costs one-quarter of the cost of a fatality. Note that these values refer to individuals killed or seriously injured and are not specific to crash costs.

203. The GDP of a country is usually reported annually by the finance ministry. If, for example, a country has a GDP of $4,800, the cost of one fatality on the roads in that country would amount to $336,000 (calculated as 70 times $4,800). The cost of a serious injury would be one-quarter of this amount, totaling $84,000.

204. The blackspot team can then calculate the total casualty costs for the location by adding up the number of fatalities at the blackspot, multiplying that number by $336,000, and then adding the number of serious casualties at the site multiplied by $84,000.

205. By dividing this total (representing the cost of all deaths and serious injuries at the site) by the number of reported crashes at the site, a reasonable figure per casualty crash can be obtained. This figure will vary from one blackspot to another, depending on the number of fatalities in some crashes. It will be skewed toward crashes involving multiple fatalities and multiple casualties. For instance, if one fatal crash involved multiple fatalities, that crash would significantly raise the casualty crash cost at the site. But until an agreed national cost for a casualty crash is developed and agreed upon, this method will suffice. As always, the accuracy of the final figure will depend on the completeness of the crash data available to the blackspot team.

> "A blackspot will only be eliminated when funding is approved, and the agreed countermeasures are implemented"

206. The next task is establishing a benchmark percentage, based on previous experience, that can demonstrate the likely crash reductions resulting from a recommended treatment. This benchmark relies on historical crash data to forecast the future safety benefits arising from a package of treatments. CRFs are used for this part of the economic assessment.

207. Developing a set of CRFs requires reliable crash data spanning several years, along with a dedicated research team capable of analyzing the data and categorizing them by countermeasure. While CRFs are explained more fully in the following section, it is important to note for now the following:

---

6   McMahon, K. and Dahdah, S. 2008. *The True Cost of Road Crashes: Valuing Life and the Cost of a Serious Injury*. International Road Assessment Programme.

- CRFs are averages from many treated blackspots.
- CRFs do change slightly over the years due to changes in the frequency of crashes at treated sites and additional follow-up analyses.
- CRFs provide a useful benchmark for comparing all blackspot treatments within a national program.

208. Box 5 provides an example of how to calculate the benefits of a package of treatments for a rural crossroad intersection in a CAREC country.

Benefit–cost ratio

209. A BCR is generated by dividing the calculated benefit of a package of countermeasures by the cost of the package. This BCR allows an equitable comparison of sites nationwide. The site with the highest BCR takes priority for funding by the national road authority, followed by the second-highest BCR, and so on, until the annual national blackspot program budget is exhausted.

210. In the worked example in Box 5, the roundabout is estimated to cost $1,420,000, while the calculated benefits (crash savings) based on CRFs are estimated to be $10,700,000.

$$\text{BCR} = \frac{\$10,700,000}{\$1,420,000}$$

$$\text{BCR} = 7.53$$

211. These BCR calculations are included as a chapter in the blackspot report, which is submitted to the national road authority or other relevant agencies to seek funding approval for the construction of the roundabout. This site will compete for funding against other blackspot sites from other parts of the country. If it does not receive funding approval this year, it will have to reapply next year or explore alternative sources of funding. A BCR of 7.53 is considered favorable in most national programs, significantly increasing the likelihood of this site and its treatment package securing funding this year.

Box 5: Calculating the Benefits of a Package of Treatments

A rural crossroad intersection experiences 15 casualty crashes in 3 years (an average of 5 casualty crashes per year). Most of the crashes are right-angle crashes, with 6 occurring at night.

In this Central Asia Regional Economic Cooperation country, the cost of a casualty crash is $107,000.

The blackspot team recommends constructing a roundabout for the site, along with installing high mast lighting on the central island. The concept cost for the roundabout is $1,420,000.

A single lane roundabout in a rural area has a crash reduction factor (CRF) of 80% for all casualty crashes. Meanwhile, street lighting has a CRF of 25% for nighttime casualty crashes, and the conceptual cost of the new lighting is $100,000.

As the roundabout is expected to achieve a larger reduction in crashes compared to the lighting, a CRF of 80% is used in the calculations. The original methodology that developed these factors allows for only one CRF per package of treatments.

The new roundabout is projected to have a lifespan of 25 years and is expected to prevent 80% of the 5 crashes per year for the entire period. As a result, it is predicted to prevent 100 casualty crashes during its lifetime.

At $107,000 per casualty crash, the roundabout is estimated to save $10,700,000 in crash costs throughout its lifetime. Therefore, the benefits from this roundabout are calculated at $10,700,000.

Source: Asian Development Bank road safety engineering consultant.

C.   Crash reduction factors

212.   A CRF is the average percentage of casualty crashes that, based on past experience, will be reduced by one treatment (such as constructing a roundabout, installing a warning sign, or using chevron alignment markers) at a blackspot. For example, converting a rural crossroad intersection into a single-lane roundabout is known to reduce casualty crashes by 80%, placing a warning sign ahead of a curve reduces casualty crashes by 25%, and installing chevron alignment markers around curves reduces casualty crashes by 30%.[7] CRFs assist the blackspot team in calculating the expected benefits of their package of recommended countermeasures.

213.   In turn, this figure is the essential numerator for calculating the BCR for a package of treatments at a blackspot. When used consistently by all road authorities across the country to calculate the BCR for their packages of blackspot treatments, it enables a simple and quick comparison among blackspots vying for national funding. This comparison allows the agency controlling the national blackspot budget to approve funding for blackspot treatment packages that offer the best economic return to the country.

> "Crash reduction factors assist the blackspot team in calculating the expected benefits of their package of recommended countermeasures"

214.   It goes without saying that obtaining such CRFs requires several years of complete before-and-after crash data. For a road authority initiating a national blackspot removal program, this may seem a significant hurdle to overcome. Most CAREC road authorities lack adequate experience with blackspot investigations and treatments to develop a country-specific or regional table of CRFs.

215.   To overcome this difficulty, especially until enough before-and-after crash data become available, CAREC countries should consider adopting an established table of CRFs. Appendix 1 is one of several CRF lists from around the world. The source, being the state road authority of Victoria (Australia), is inconsequential, as the individual figures are less important than the relative figures and the consistency they offer within a blackspot removal program. An agreed-upon set of CRFs simply allows every team investigating blackspots across the country to use the same set of factors. In turn, this approach leads to quantifiable benefits resulting from a package of blackspot treatments assessed uniformly using the same CRFs as every other blackspot in the country.

216.   In time, it is expected that road safety engineers and university researchers across CAREC will evaluate the performance of various blackspot countermeasures and develop a table of CRFs specifically tailored for use in CAREC.

D.   Writing a blackspot report

217.   Once a blackspot has been identified and thoroughly investigated, crash patterns (if any) identified, and a package of practical countermeasures developed after on-site inspections, the blackspot team compiles a comprehensive report. This blackspot report serves as both a record for future reference and a formal request to the national authority for funding approval for the recommended countermeasures. Funding for one site is unlikely to be approved until the managers of the national blackspot removal program are satisfied that the BCR calculated for the site is at least comparable to, if not higher than, that of other blackspots across the country. To determine this, the managers need to review all the blackspot reports submitted from across the country for that year, enabling them to compare the BCRs.

218.   Each blackspot report should explain the nature of the crash problem at the site, justify the proposed countermeasures, and provide a logical basis for funding approval. A blackspot report typically consists of eight sections:

- **Title page.** This page provides information about the location/name of the blackspot, the client for whom the investigation has been undertaken, the names of the blackspot team members, and the date of the report.

---

[7]   VicRoads (the State Road Authority of Victoria) Crash Reduction Factors. 2017. Melbourne.

- **Site description.** This section offers a detailed description of the location, including the road name and/or road number, kilometer position (start and finish kilometers of a road section), intersecting road name (if any), town, and district names.
- **Existing conditions.** This part provides a description of the site, including details about the terrain and any abutting development. It includes information about traffic volumes, pedestrian activity, and the operating speed environment if known or as estimated. An aerial photograph of the site can also be added.
- **Analysis of crash data.** This part summarizes all reported crashes at the site, covering the last 3 years if possible. It also includes the collision diagram and the crash factor matrix used for this site. Any relevant previous blackspot reports can be referenced here.
- **Contributing factors.** In this section, the blackspot team assesses the site conditions contributing to the crash patterns at the blackspot. If possible, photos of the site showing the specific contributing factors should be added.
- **Countermeasures.** This section lists the proposed countermeasures for the site, plus a concept design for the whole package. Details of multiple options being considered (such as high-cost and low-cost options, or interim improvements pending construction of a final recommended countermeasure) should be provided.
- **Economic assessment.** This section itemizes the concept cost estimate for all the countermeasures. It also shows the CRF used for the recommended countermeasure along with the calculated benefits (in $) and then it presents the BCR for the option.
- **Appendices.** This section may include newspaper articles about crashes at the site, statements from local residents/farmers, more site photographs, or a summary of earlier blackspot reports about the site.

"Blackspot reports offer a concise summary of the investigation and a detailed economic assessment leading to the benefit–cost ratio"

219. Eventually, the national road authority may prepare a standard template to be used by all districts when reporting their blackspot investigations and seeking funding approval.

E. Interim countermeasure treatments

220. Many blackspots can be eliminated by installing signs, delineation, and line marking. These measures are typically low-cost solutions and, when designed and implemented by experienced engineers and contractors, prove to be effective in reducing the number of future crashes.

221. However, some blackspots may require a more substantial package of treatments. If the recommended treatments need to be designed, and especially if they entail land acquisition or service relocation, there may be a significant delay before the work can be executed. This will postpone the predicted crash reductions and the subsequent community savings from reduced road trauma.

222. In such situations, the blackspot investigation team, in collaboration with the responsible road officers, may consider breaking down the recommended treatments into staged works. Stage 1 works typically consist of low-cost, quickly installed countermeasures that provide interim safety improvements and immediate safety benefits. These may include warning signs, line markings, speed restrictions, or other minor improvements designed to control, warn, or guide drivers. In some cases, small-scale civil works, such as splitter islands at intersections or paved shoulders, could be advanced and implemented in Stage 1.

223. Meanwhile, the more extensive Stage 2 works, such as an intersection reconstruction, can be programmed for the following year, with the expectation that some improvement in crash performance at the site is already underway.

224. Importantly, once a blackspot investigation has determined the treatment options, it is advisable to program the approved works as early as possible to commence crash savings.

Box 6: Visual Deceit

Visual deceit is the term used to describe situations in which a combination of factors, such as a curve in the road, proximity to a line of power poles, and a long straight hedge row, can lead some drivers to believe the road continues in one direction when it actually proceeds in another. While engineers routinely use line marking, guideposts, and signs to inform drivers of the correct path, visual deceit can still occur when other roadside features dominate the landscape. Visual deceit can mislead some drivers, and it is a factor that a blackspot team should carefully consider, especially at high-speed rural locations. Stronger delineation may reduce visual deceit, but it is not always a guaranteed solution as only a very small percentage of drivers may be affected by the problem. In many cases, more extensive changes to the environment, such as relocating the intersection, undergrounding power lines, or closing the side road, may be needed.

Source: Asian Development Bank.

Visual deceit on a right-side road. If a driver on a highway in a right-side driving country mistakes the route due to the presence of power poles and a hedge, thinking the highway goes straight ahead, they may end up on an unpaved side road with minimal risk of serious incident.

Visual deceit on a left-side road. If a driver on a highway in a left-side driving country mistakes the route due to the presence of power poles and a hedge, thinking the highway goes straight ahead, and a vehicle is traveling around the curve from the opposite direction at the same time, it can lead to a serious high-speed head-on collision.

Box 7: See-Through Issues

See-through issues arise when a driver's attention is drawn to a distant object, diverting their focus from immediate concerns, such as a nearby intersection, important signs, or pedestrians at a crossing.

When a blackspot investigation team suspects a see-through issue at a site, finding the countermeasures is not always straightforward. Blocking off the distant item is rarely possible, so the focus shifts to increasing the conspicuity of nearby items, such as crossings, intersections, or signs. This requires careful consideration of sign selection and placement, although it may not always guarantee a resolution to these issues.

Source: Asian Development Bank road safety engineering consultant.

225. These blackspot case studies are based on sites along CAREC roads. In most examples, crash data were limited or nonexistent. These case studies have been included in this manual to provide practical examples of, and lessons about, the blackspot process. Crash data have been added where necessary.

226. Case studies from various road types and areas (urban and rural) are featured in this chapter to provide several examples of blackspots. These case studies showcase the blackspot team's recommended actions to address distinct crash patterns at these sites. They are illustrative of real blackspot investigations and are presented in an abbreviated format, condensing a full blackspot report to save space while emphasizing the reasoning of the investigation team and the importance of the BCR in the blackspot process.

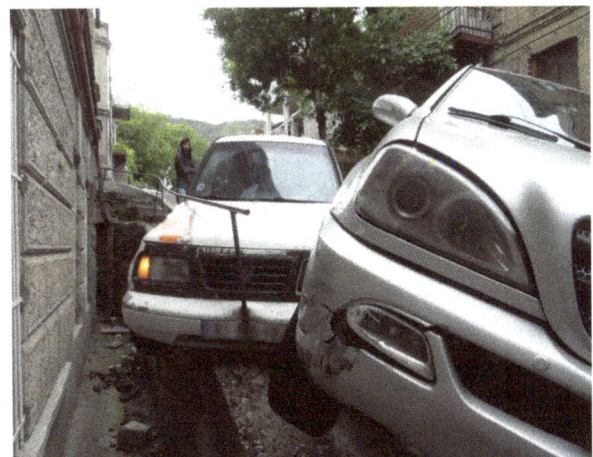

Many types of road crashes. Some involve a single vehicle, others include multiple vehicles. All involve trauma to the people involved. Removing blackspots is a cost-effective way of minimizing this trauma and reducing the cost of road crashes.

CASE STUDY 1   A blackspot at an urban crossroad intersection

An urban arterial crossroad blackspot. This intersection has straight approach roads and is surrounded by residential development. It has been the location of 12 casualty crashes in the past 3 years.

Blackspot at an urban arterial crossroad intersection

Site description

An intersection of two urban arterial roads has witnessed 12 casualty crashes over the past 3 years, mostly occurring during dusk and nighttime. The intersection is flat, controlled by give-way signs on the north and south approaches. Both roads are straight with two traffic lanes plus parking. Vehicle speeds often exceed the 60 km/h urban speed limit. The intersection is surrounded by residential housing, plus a small corner shop.

Crash patterns

Right-angle collisions involving vehicles traveling south and east (7 crashes) Right-angle collisions involving vehicles traveling south and west (2 crashes)	Collision diagram (below)
Eleven of the 12 reported crashes happened during darkness, or dusk	Crash factor matrix (below)

Contributing factors	Crashes with common factors		
Lack of intersection conspicuity Small and distant placement of give-way signs Limited sight distance	9		
Wide roads			
Suspected overshoot problem	9		
High speeds (60+ km/h), especially at night	3		

Possible countermeasures	Will this address main problems?	Estimated cost	CRF(%)	Life years	BCR
Construct a single lane roundabout	Yes	$1,450,000	70	25	5.2
Install street lighting	Partially	$40,000	40	20	-
Duplicate give-way signs on the north and south approaches	Partially	$5,000	15	15	-

Crash reduction factor (CRF): 70% (roundabout)
Predicted crash reduction: 70% of 12/3 crashes per year = 2.8 crashes per year
Life of roundabout: 25 years
Total crashes predicted to be prevented: 70 crashes
Benefit calculation: 70 crashes x $110,000 (the national average casualty crash cost) = $7,700,000
Cost of works = $1,490,000 (including the roundabout and lighting)
BENEFIT–COST RATIO (BCR): 5.2

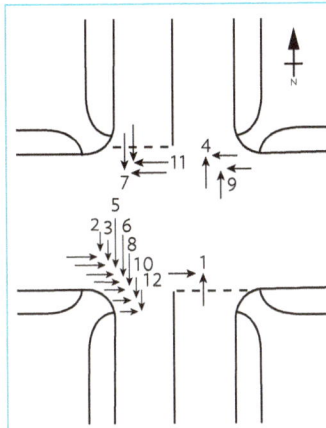

Crash Number	1	2	3	4	5	6	7	8	9	10	11	12
Date: month	13/07	04/08	08/06	03/07	07/11	27/02	03/05	24/07	18/04	21/05	14/06	20/08
Day of week	Sat	Wed	Sun	Thurs	Fri	Fri	Sun	Fri	Sun	Fri	Mon	Fri
Time of day	18.00	18.55	19.00	22.45	21.45	18.20	18.30	20.00	18.45	16.10	20.35	18.15
Severity	3	1	3	2	1	3	2	2	3	2	2	3
Light conditions												
Road conditions	Wet	Wet	Dry	Dry	Dry	Dry	Dry	Dry	Dry	Dry	Wet	Dry
DCC code	110	110	110	110	110	110	110	110	110	110	110	110
Object 1	Car	Car	Car	Car	Car	Car	Car	Car	Car	Car	Van	Car
Object 2	Car	Car	Car	Car	Car	Van	Car	Car	Car	Car	Car	Car
Direction 1	N	S	S	N	S	S	S	S	N	S	S	S
Direction 2	E	E	E	W	E	E	W	E	W	E	W	E
Other												

The collision diagram shows that most of the right-angle collisions involve vehicles from the north striking vehicles from the west.

The crash factor matrix illustrates that the majority of the crashes occur during dusk or at night.

DCC = description for classifying crashes, E = east, N = north, S = south, W = west.

Discussion

227. The blackspot team attributes the right-angle collisions to an overshoot problem, which is compounded by the intersection's lack of conspicuity from the north and the high approach speeds.

228. Treatment options to reduce the likelihood of crashes include traffic signals, a roundabout, additional or duplicate give-way signs (or conversion to stop signs), street lighting, and channelization. In line with Safe System principles, a single-lane roundabout is recommended; it is one of the few treatments proven to mitigate overshoot and restart problems.

229. Street lighting will help reduce the likelihood of a crash, but it will not impact crash severity. Four additional streetlights are part of the package. The single largest CRF is for a single-lane roundabout (70%). For specific CRFs, please refer to Appendix 1.

230. Here are the three main factors affecting road safety at this urban crossroad intersection blackspot, along with possible countermeasures:

231. Factors influencing the likelihood of a crash:

- Constructing a roundabout
- Duplicating stop signs

- Installing street lighting to make the intersection more conspicuous at night

232. Factors influencing exposure to a crash:

- There are no feasible options for diverting traffic from this intersection. The closure of one or more approaches is not a viable solution. Right in/ right out channelization is impractical in this context.

233. Factors influencing the severity of a crash:

- Constructing a roundabout
- Ensuring consistent and safe speed management

Outcome

234. Duplicate stop signs (one on each side of the road) were installed on both the north and the south approaches, line marking at the intersection was renewed, and two new streetlights were installed as part of the Stage 1 treatment package. During the first 12 months of implementing these treatments, reported casualty crashes decreased by 35%. While the design for a single-lane roundabout continues, the road authority has postponed a decision on constructing the roundabout pending ongoing crash site monitoring.

CASE STUDY 2   A blackspot involving many pedestrian collisions on a highway passing through a village

A blacklength through a village. This village has a road safety problem extending over a length of almost 1 kilometer. It involves crashes at side road intersections and pedestrian collisions.

Blackspot involving many pedestrian collisions on a highway passing through a village					
Site description					
A two-lane undivided highway passes through a large rural settlement. The highway is flat and quite straight. The road reserve is wide. Many pedestrians and motorcyclists use the highway in the village, while cars, trucks, and buses pass through the village. Despite the posted 60 km/h speed limit, many of these vehicles exceed this speed, especially at night.					
Crash patterns					
Right-angle collisions at side road intersections (4 crashes) Rear-end and head-on collisions, mid-block (5 crashes) Pedestrian incidents (6 crashes)			Collision diagram (see below)		
11 of the 15 crashes happened during darkness or dusk Dark (crashes 1, 2, 3, 6, 8, 10, 12, 14, 15) Dusk (4, 7) Remaining (4) crashes are daytime			Crash factor matrix (see below)		
Contributing factors	**Crashes with common factors**				
Wide highway with no median, worn lines, and few signs	4				
Lack of pedestrian facilities	6				
High speeds, especially at night	12				
Unsafe intersection layouts	3				
Inadequate street lighting	8				
Possible countermeasures	**Will this address main problems?**	**Estimated cost**	**CRF(%)**	**Life years**	**BCR**
Traffic management: road humps, gateway signs	Partially	$200,000	20	20	
Pedestrian refuges for spatial separation and as a median for traffic	Partially	$200,000	50 pedestrian crashes 67 fatal pedestrian crashes	25	

Installation of 3 raised pedestrian crossings	Partially	$125,000	63	20	9.1
Implementation of new street lighting	Partially	$120,000	50 night crashes only	20	
Installation of warning and speed restriction signs	Limited	$30,000	15	15	

Crash reduction factor (CRF): 63% (raised crossings)
Predicted crash reduction: 63% of 15/3 crashes per year = 3.15 crashes per year
Life of humped crossings: 20 years
Total crashes predicted to be prevented = 63 crashes
Benefit = 63 x $98,000 (the national average casualty crash cost) = $6,174,000
Cost of works = $675,000 (includes traffic calming, pedestrian refuges, humped crossings, and new lighting)
BENEFIT–COST RATIO = 9.1

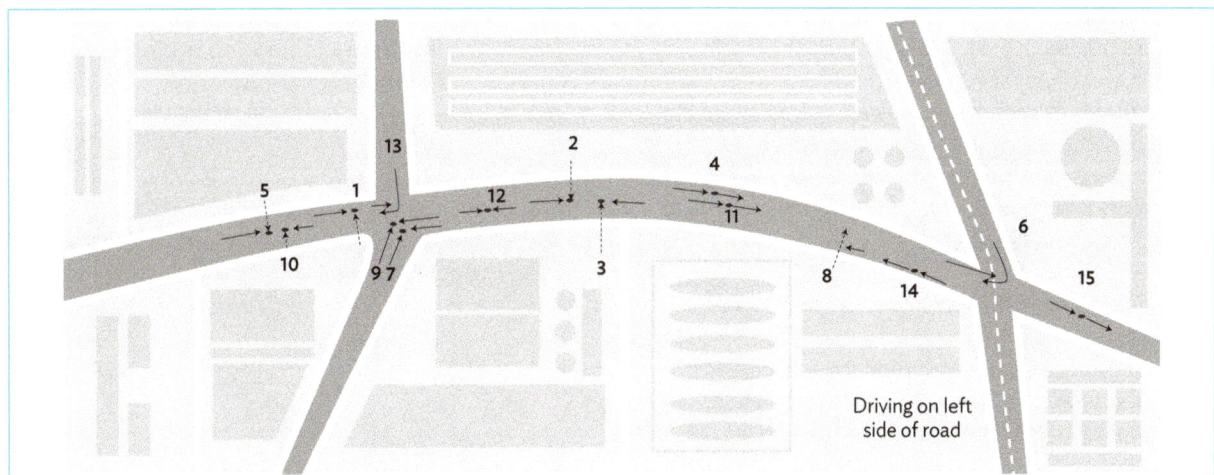

The collision diagram shows a pattern of pedestrian collisions throughout the village, right-angle collisions at side roads, and rear-end collisions near the center of the village.

Crash Number	1	2	3	4	5	6	7	8	9	10	11	12	13	14	15
Date: month	2/03	5/05	11/10	29/11	20/01	28/03	1/04	5/09	8/12	31/12	2/02	10/03	5/06	7/09	2/12
Day of week	Sun	Fri	Wed	Wed	Sat	Wed	Sun	Wed	Sat	Mon	Mon	Sun	Wed	Sat	Sun
Time of day	01.15	22.30	19.50	17.50	11.10	20.55	18.30	23.00	14.40	04.00	07.45	23.30	11.00	20.30	21.30
Severity	1	2	2	2	3	3	2	1	2	1	2	1	3	2	2
Light conditions															
Road conditions	Wet	Dry	Dry	Dry	Dry	Dry	Wet	Dry	Dry	Dry	Dry	Dry	Dry	Dry	Dry
DCC code	102	100	100	130	100	111	110	100	110	100	130	120	111	130	130
Vehicle 1	Car	Car	Bus	Bus	Car	Car	M/C	Car	M/C	Car	M/C	M/C	Car	Car	M/C
Vehicle 2	Ped	Ped	Ped	Truck	Ped	Truck	Car	Ped	Car	Ped	Truck	Car	Truck	Car	Car
Direction veh. 1	E	E	W	E	E	E	N	W	N	W	E	E	S/W	W	E
Direction veh. 2	N	S	N	E	S	S/W	W	N	W	N	E	W	E	W	E

DCC = description for classifying crashes, E = east, M/C = motorcyclist, N = north, S = south, S/W = southwest, W = west.

235. The crash factor matrix shows that the majority of the crashes occur during dusk or at night, with some happening in rainy conditions.

Discussion

236. The blackspot team inspected the highway through the village during both the day and the night, observing many vulnerable road users, including pedestrians, bicyclists, handcarts, and motorcyclists. They found inadequate street lighting, a lack of pedestrian facilities, and poor traffic control at side road intersections. They noted instances of high traffic speeds, with several vehicles surpassing the posted 60 km/h speed limit. Speeds were noticeably higher at night as traffic volumes decreased. In light of these findings, the blackspot team considered various treatments designed to reduce the likelihood of fatal and serious casualty crashes:

237. Influencing the likelihood of a crash:

- Setting and enforcing appropriate speed limits
- Implementing traffic calming measures, such as gateway signs, humped crossings, and refuge islands
- Constructing pedestrian refuges interconnected by line marking to create a central median along the highway
- Improving traffic control (e.g., duplicating give way signs) at side roads

238. Influencing the exposure to a crash:

- There is no practical option to take to redirect the traffic passing through the village onto a bypass route.
- Constructing footpaths was considered but eventually rejected as the pedestrian safety issue involves crossing the highway.
- The refuge islands will create a central median, reducing head-on crash exposure.

239. Influencing the severity of a crash:

- Ensuring consistent and safe speed management

Outcome

240. The blackspot team provided the following recommendations:

- Placing a pair of large gateway signs at each end of the village, each featuring a 40 km/h speed restriction sign and an additional warning plate "HUMPS AHEAD" underneath.
- Installing five pedestrian refuges—2.5 meters (m) wide and 30 m long—along a 500 m section of the highway, linked together with line marking.
- Introducing three humped pedestrian crossings at locations with the highest pedestrian demand.
- Implementing three additional road humps at agreed locations near side roads.
- Adding new streetlights through the village.

241. Funding for these works was approved by the road authority based on a high BCR of 9.1. Designs have commenced, and the work is expected to take place in the coming months.

CASE STUDY 3   A rural blacklength within a series of curves on a highway in a hilly area

A series of sharp curves on a rural highway. Unexpectedly sharp curves on highways, usually in high-speed sections of rural highways, are often the sites of run-off-road crashes. This type of crash can cause severe injuries, but there are low-cost countermeasures that can be applied to prevent these crashes.

A rural blacklength within a series of curves on a highway in a hilly area.

Description of blacklength

A highway traverses hilly terrain in a rural area. It has numerous sharp curves, with the two tightest near the highest point. The road includes long, steep grades (8%–9%), combined with these sharp curves. Traffic volumes on this route are low, with speeds varying from 40 km/h (trucks, buses) to 80 km/h (cars). Some drivers tend to enter these two curves too fast. If they respond too late while navigating either curve, the risk of leaving the road is high. In case a vehicle departs from the road, the crash consequences are severe due to inadequate safety barriers.

Crash patterns	
Run-off-road (ROR) crashes involving vehicles from the south (6 crashes) ROR crashes involving vehicles from the north (2 crashes)	Collision diagram (below)
3 of the 8 reported crashes happened during darkness, and another 2 occurred in fog	Crash factor matrix (below)

Contributing factors	Crashes with common factors		
Long steep grades: some trucks and buses lose brakes	3		
Cars travel too fast for the abrupt changes in geometry	4		
The tight curves are in an isolated part of the highway and are the sharpest curves, in the steepest section of this road	8		
Delineation is worn, and previously installed safety barriers have been damaged but not repaired	3		
Inadequate signs, including warning and speed restriction signs	3		

Possible countermeasures	Will this address main problems?	Estimated cost	CRF(%)	Life years	BCR
Widen the road through the curves and pave shoulders (2 m wide)	Partially	$1,500,000	35	25	
Implement a wide centerline (painted median) with tactile line markings, raised reflective pavement markers (RRPMs), and chevrons with road widening	Partially	$1,545,000	60	15	3.8

Install 850 m of safety barriers	Partially	$95,000	0	20	
Delineate the route with tactile edge lines and centerlines through the 1.5 km blacklength	Partially	$25,000	23	5	
Delineate sharp curves with reflective chevron alignment markers	Partially	$20,000	30	15	
Install large advance warning signs at both ends of the highway where the hill section begins	Limited	$10,000			

Crash reduction factor (CRF): 60% (wide centerline [painted median] with tactile line markings, RRPMs, and chevrons on rural undivided roads with or without road widening)
Predicated crash reduction: 60% of 8/3 crashes per year = 1.6 crashes per year
Life of treatment: 15 years
Total crashes predicted to be prevented: 24 crashes
Benefit: 24 x $265,000 (the national average casualty crash cost) = $6,360,000
Cost of works = $1,650,000 (includes tactile edge lines, widening, and delineation)
BENEFIT–COST RATIO = 3.8

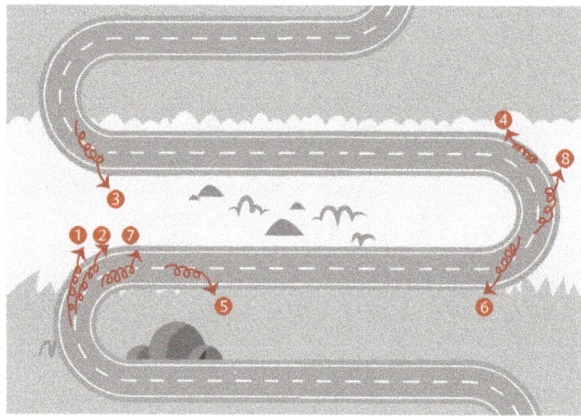

Collision Diagram

Crash Factor Matrix

Crash Number	1	2	3	4	5	6	7	8
Date: month	6/03	1/05	21/08	28/02	28/05	1/07	5/08	8/10
Day of week	Sun	Fri	Sat	Sat	Wed	Sun	Fri	Sat
Time of day	07.30	23.30	10.30	06.50	23.00	18.30	22.00	14.40
Severity	2	1	2	2	1	2	2	1
Light conditions	Fog			Fog				
Road conditions	Wet	Dry	Dry	Wet	Dry	Wet	Dry	Dry
DCC code	180	180	182	182	184	180	180	182
Vehicle 1	Car	Bus	Truck	Bus	Car	Car	Truck	Car
Vehicle 2								
Direction veh.1	N	N	S	N	E	S	N	N
Direction veh.2								
Observations			Brake Failure					

DCC = description for classifying crashes, E = east, N = north, S = south.

Discussion

242. The blacklength has been investigated several times over the last 10 years. It is in a high-speed area, and the blackspot team is well aware of the heightened risk of fatal and serious injuries resulting from any crashes in this location. Earlier attempts at improving delineation and installing safety barrier have not yet yielded the anticipated results. Consequently, the blackspot team has now considered implementing three elements of Safe System-compliant crash countermeasures:

243. Influencing the likelihood of a crash:

- Benching (cutting back) the inside of the two sharp curves, widening the curves and shoulders, and paving the shoulders on both sides throughout the whole 1.5 kilometers (km) blacklength
- Better, more conspicuous warning and advisory speed signs
- Implementing tactile edge lines along the blacklength and tactile centerlines (a solid "no overtaking" line) within the curves
- Installing reflective guideposts on both sides of the highway through the blacklength
- Placing oversized, reflective chevron alignment markers around both curves

244. Influencing the exposure to a crash:

- Redirecting traffic away from the sharp curves is not a feasible option due to the challenging terrain and the high cost associated with altering the road alignment

245. Influencing the severity of a crash:

- Installing appropriately designed safety barriers
- Managing and enforcing Safe System-compliant speeds

Outcome

246. Funding for the package of countermeasures recommended by the blackspot team has been approved by the road authority, and the work is scheduled for the next spring/summer.

CASE STUDY 4   A rural blackspot at a complex intersection

Y-junctions are high-risk locations. Removing Y-junctions, particularly those in high-speed areas, is one of the most useful road safety activities a road authority can undertake. Y-junctions have high angles of impact and high relative impact speeds between conflicting directions of travel.

Blackspot at a complex rural intersection on a CAREC highway

Site description	
A major rural road intersects a CAREC highway within a large sweeping curve. Two roads lead from the highway to the rural road, following a typical design from many years ago when traffic volumes were very low. Both roads are two-lane, with wide but unpaved shoulders. The site is flat, and visibility in each direction is adequate. All line markings are worn, and there is no other delineation. Small give way signs are present on the side roads, and direction signs can be found on both approaches to the minor road from the highway. A single streetlight is near the service station, about 100 m from the intersection.	

Crash patterns	
Head-on collisions involving vehicles turning to/from the highway at three Y-junctions	Collision diagram (below)
11 of the 12 reported crashes occurred during darkness or dawn	Crash factor matrix (below)

Contributing factors	Crashes with common factors		
Wide highway with high traffic volumes	2		
Unsafe geometry	12		
Inadequate traffic control	4		
High speeds on the highway (100+ km/h), especially at night	4		

Possible countermeasures	Will this address main problems?	Estimated cost	CRF(%)	Life years	BCR
Construct a single-lane roundabout	Yes	$2,900,000	70	25	
Convert the Y-junctions into a single T-junction	Yes	$2,600,000	85	25	5.6
Install street lighting	Partially	$90,000	40	20	
Implement new signs and line marking	Partially	$30,000	15	15	

CRF: 85% (removal of Y-junctions)
Predicted crash reduction: 85% of 12/3 crashes per year = 3.4 crashes per year
Life of Y-junction removal: 25 years
Total crashes predicted to be prevented: 85 crashes
Benefit: 85 x $180,000 (the national average casualty crash cost) = $15,300,000
Cost of works: $2,720,000 (includes removal of Y-junctions with either a roundabout or a T-junction, plus signs, lines, and lighting)
BENEFIT–COST RATIO = 5.6

Collision Diagram

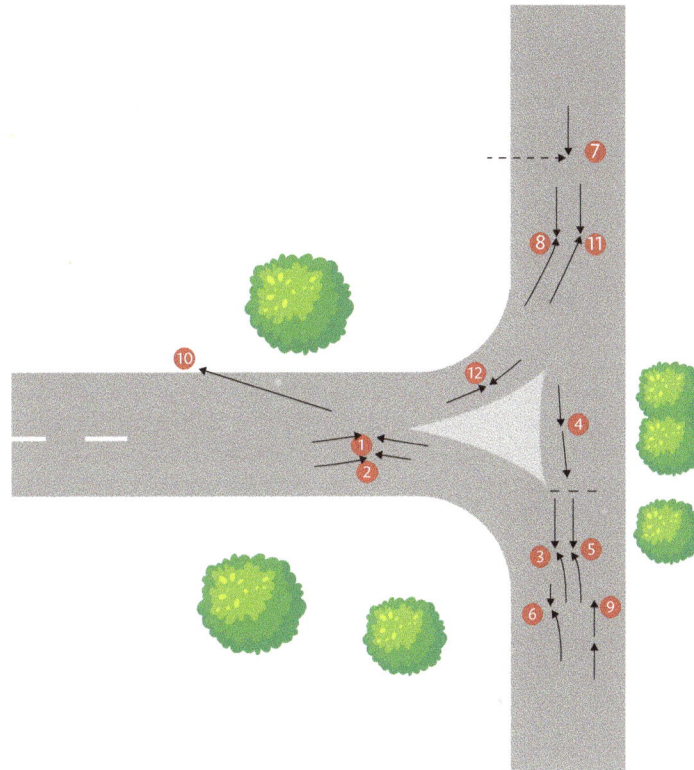

Collision Diagram for the Intersection

Crash Factor Matrix

Crash Number	1	2	3	4	5	6	7	8	9	10	11	12
Date: month	12/03	14/05	11/07	29/01	28/03	1/04	5/09	8/02	31/04	26/06	10/08	7/09
Day of week	Sun	Fri	Wed	Wed	Wed	Sun	Wed	Sat	Mon	Tues	Sun	Sat
Time of day	13.00	23.30	20.30	16.50	23.00	18.30	22.00	17.40	04.00	04.00	23.30	20.30
Severity	1	2	2	3	1	2	2	1	1	2	1	3
Light conditions												
Road conditions	Wet	Dry	Dry	Dry	Dry	Wet	Dry	Wet	Dry	Wet	Dry	Dry
DCC code	122	122	122	130	122	122	100	120	130	182	122	120
Vehicle 1	Car	Car	Bus	Bus	Car	M/C	Car	Car	Car	Truck	M/C	Car
Vehicle 2	Bus	Truck	Truck	Car	M/C	Bus	Ped	Car	M/C	?	Truck	Car
Direction veh.1	E	E	S	S	S	S	S	E	N	N/W	E	E
Direction veh.2	N	N	N/W	S	N/W	N/W	E	S	N	?	S	W
Observations			Speed	Speed							Speed	

DCC = description for classifying crashes, E = east, M/C = motorcyclist, N = north, S = south, N/W = northwest.

Discussion

247.   The blackspot team is well aware that Y-junctions are notoriously unsafe intersections, especially in high-speed areas, posing a very high risk of FSI crashes. Recognizing the urgency of this issue, the blackspot team has decided to alter the intersection and remove the Y-junctions. They know this process will take time due to various steps like agreeing on the new layout, design, land acquisition, and construction.

248.   In response to the high crash risk, the team has considered three groups of crash countermeasures for this location. They understand that immediate countermeasures are needed within a staged treatment program. Low-cost treatments are considered for immediate installation, and longer-term treatments are developed as the design and land acquisition phases are completed.

249.   Influencing the likelihood of a crash:

- Roundabout
- T-junction
- Closure of side roads
- Duplicating the give way signs and adding pavement markings
- Lighting
- Setting and enforcing reduced speed limits

250.   Influencing the exposure to a crash:

- Building an alternative route to divert traffic from this intersection is not practical. Closing the highway is not an option.
- Reducing overshoot crashes can only be achieved by replacing the Y-junctions with a roundabout or a T-junction and channelization.

251.   Influencing the severity of a crash:

- Roundabout
- T-junction and channelization
- Consistent and safe speed management

252.   Because of the high crash severity at this site, the highway authority quickly decided to introduce a package of short-term, low-cost treatments, including:

- duplicating the existing give-way signs to minimize overshoot risk,
- renewing all line marking, including edge lines,
- installing six new streetlights, and
- removing the blue/white bifurcation sign on the central island.

253.   The authority also recognized that such intersection layouts, which may have been acceptable when traffic volumes were low, are no longer safe with the increased traffic on CAREC highways today. As a result, they have resolved to:

- gather data on traffic volumes and land acquisition availability and costs to decide between a roundabout or a T-junction for this site. While a roundabout is a safe form of intersection under the Safe System approach, the final decision will be based on a cost-benefit analysis.

Outcome

- Duplicate give way signs, new line markings, and streetlights were promptly installed by the road authority.
- Additional traffic direction signs were installed on each approach.
- Traffic counts were conducted, and discussions commenced about the type of intersection to replace the Y-junctions. An elongated central island design for a roundabout is the preferred option and is expected to be built in the following year.
- Preliminary crash reports for the first 9 months of operation with the new treatments indicate a nearly 30% decrease in head-on crashes (due to overshoot).

CASE STUDY 5   Mass action treatments

Mass action treatments can greatly improve long lengths of highway. When crash data are incomplete, and budgets are limited, it can be a useful and practical initiative to apply proven low-cost treatments along a route to counter particular crash types.

Mass actions along a rural section of a CAREC highway					
Route description					
A rural CAREC highway, linking the capital city with a major provincial town, spans 120 km. The road has a history of head-on and run-off-road (ROR) collisions, distributed along the route, with over 107 crashes reported in the past 3 years. The highway has two lanes, with wide but unpaved shoulders, and passes through flat and rolling terrain. The line marking is worn, with no other delineation.		Typical photos (above)			
Crash patterns					
Head-on collisions (at least 35 FSI head-on crashes in the past 3 years)		Collision diagram (not needed)			
ROR collisions (at least 72 FSI ROR crashes in the past 3 years)					
At least 30 crashes occurred during darkness		Crash chart (below)			
Contributing factors	**Crashes with common factors**				
High traffic volumes	12				
Unpaved shoulders	12				
Few overtaking lanes	35				
Inadequate delineation	30				
High speeds on the highway (100+ km/h), especially at night	40				
Possible mass action treatments	**Will this address main problems?**	**Estimated cost**	**CRF(%)**	**Life years**	**BCR**
Tactile edge lines only	Partially	$480,000	23	5	
Tactile center lines	Partially	$240,000	80 head-on crashes 60 ROR to the left	5	6.3
Paved shoulders (only at curves) and tactile edge lines	Partially	$1,000,000	30 of ROR crashes 14 fatal/serious crashes	25	

| Delineation of curves (chevron alignment markers) | Partially | $60,000 | 30 ROR crashes on curves | 15 | |
| Reflective guideposts | Partially | $80,000 | 7 | 15 | |

CRF: 80% (of head-on crashes)
Crash reduction used for tactile center lines: 80% of 35 head-on crashes in 3 years = 9.3 crashes per year
Life of tactile center line: 5 years
Total crashes predicted to be prevented: 47 head-on crashes
Benefit: 47 x $250,000 (the national average casualty crash cost) = $11,666,000
Cost of works = $1,860,000 (includes tactile line marking plus signs)
BENEFIT–COST RATIO = 6.3

Crash Chart

DCC Code	120	120	170	170	172	172	Total
Crash Type	Head-On	Head-On	Off-Road to Left	Off-Road to Left	Off-Road to Right	Off-Road to Right	-
Crashes	10	25	12	20	18	22	107
Severity (FSI)	3/7	12/13	4/8	5/15	6/12	6/16	36/71
Light conditions							
% Wet roads	20	12	9	10	12	14	12
% Northbound	-	-	50	40	33	45	-

Discussion

254. Faced with a long highway with scattered crashes, the blackspot team opts for a mass action approach to the treatments. The most serious and frequent crashes involve high-speed head-on collisions and ROR incidents. The team adheres to the Safe System approach, focusing on reducing FSI crashes by implementing proven treatments. They propose using tactile edge lines and tactile centerlines as the primary treatment for the entire highway. In addition, to increase nighttime delineation, they propose the consistent installation of reflective guideposts on both sides of the highway, both on straight segments and curves, as well as reflective chevron alignment markers for the most critical curves. They also plan to duplicate speed restriction signs and install more reminder speed restriction signs.

255. The team considers the three groups of crash countermeasures:

256. Influencing the likelihood of a crash:

- Tactile edge lines
- Paved shoulders
- Tactile center lines
- Setting and enforcing signed speed limits
- Consistent delineation with chevron alignment markers and reflective guideposts

257. Influencing the exposure to a crash:

- Currently, there are no practical options to divide this highway (with a median) or build an alternative route for diverting traffic. Closing the highway is not feasible.
- Constructing overtaking lanes, while effective in reducing head-on collisions, was discounted due to cost. This will be pursued as a separate, progressive, and long-term project.

258. Influencing the severity of a crash:

- Installing safety barriers to shield against significant roadside hazards
- Ensuring consistent and safe speed management

Outcome

259. The tactile line marking (edge lines and centerlines), reflective guideposts, and chevron alignment markers were successfully installed. However, shoulder sealing was deferred due to budget constraints. It is earmarked for future implementation when additional funds become available.

260. Initial crash reports for the first 18 months of operation with the new treatments indicate a 40% decrease in both head-on and ROR crashes.

VIII. The Next Steps

261. This manual outlines the blackspot process to guide and assist CAREC road authorities and others in investigating hazardous locations in a logical and detailed manner. The emphasis is on proven and cost-effective treatments to reduce the likelihood of a fatal or serious casualty crash. Various topics are covered by the manual, such as the following:

- The role of engineers in ensuring safe roads
- The chain of events leading to a crash
- The Safe System approach to safety
- The benefits of complete crash data
- How to identify blackspots
- How to investigate blackspots, including tools for finding crash patterns
- Treating blackspots with effective countermeasures
- Determining funding priorities for blackspot treatments

262. The manual also encourages those who manage the road network to establish a national blackspot removal program. Given the many blackspots across CAREC's roads and highways, their removal will take many years of dedicated work. The sooner that removal work starts, the safer the road network will be. National blackspot removal programs are a great investment in road safety, providing ongoing support, resources, and assurance that the blackspot team's efforts will continue into the future.

A. Establishing a national or provincial blackspot removal program

263. Reducing road trauma in CAREC countries requires stakeholders working together in a coordinated manner, guided by an agreed-upon national road safety strategy. Governments across CAREC are strongly encouraged to develop, fund, and implement a national road safety strategy as an essential step toward enhancing road safety effectively and sustainably.

264. The treatment of high-frequency crash locations constitutes just one part of a national road safety strategy, yet it is an important and cost-effective one.

A comprehensive program to identify, investigate, and treat high-frequency crash locations should be a high priority for national road authorities within the national strategy framework.

> "A blackspot removal program is a positive safety investment that yields significant dividends for a country"

265. When establishing such programs, it is necessary to focus on fatal and serious crash outcomes and demonstrate that the problem can be addressed cost-effectively. This requires an understanding of the safety benefits that a focused program of road infrastructure improvements can deliver. With this knowledge, a compelling case can be made for securing the necessary funding for safe road infrastructure.

266. As part of a national blackspot removal program, a systematic process needs to be established to identify high-frequency crash sites, investigate these sites, select effective treatments, prioritize these treatments, and implement them accordingly. In the post-installation period, the process requires program monitoring and evaluation.

267. Several institutional arrangements should be introduced to support a national blackspot removal program. The first is the availability of good-quality crash data, requiring close cooperation between traffic police and the road authority. A national road safety strategy should emphasize the importance of shared crash data (excluding personal details) between traffic police and key stakeholders as a foundational element of all national road safety programs.

268. The national road safety strategy should also stress the importance of skilled road safety engineers. A national blackspot removal program requires well-

trained technical staff, primarily engineers in the road authority, to undertake the investigations, although consultancies could also be involved. This may require hiring more road safety engineers in national road authorities, as the current count of these specialists in CAREC is low. This number should increase to fully serve national blackspot removal programs on the scale required by most countries.

269. Additionally, the road safety strategy should highlight an annual budget for blackspot removal works. Without adequate funding, the recommended countermeasures cannot be installed. The blackspots will then remain untreated, leading to unfulfilled expectations of the blackspot process. Funding is a crucial aspect, and blackspot programs worldwide have shown positive returns for the community. A common return of $4 in crash savings for every $1 spent on the site is quoted by one road authority.[8] Adequate blackspot funding should be a priority in CAREC countries at a level commensurate with the total cost of FSI in the country. Even small funding levels can be highly beneficial, provided that the funding is guaranteed annually in line with the national strategy.

270. CAREC road authorities are, therefore, urged to prioritize road safety and address four key aspects:

- Assist in the development of a national road safety strategy, including actions to identify, investigate, and treat blackspots.
- Commit to a nationwide blackspot removal program.
- Collaborate closely with traffic police to share crash data analysis, benefiting all stakeholders.
- Monitor the performance of treated blackspots and take early action if the treatments do not reduce crashes as expected.

B. Monitoring treated blackspots

271. Post-implementation monitoring of treated blackspots is the final step in the blackspot process and holds significance for several reasons.

272. First and foremost, the blackspot team should pay close attention to a site immediately after treatment in case things go badly wrong. Road authorities bear the responsibility of ensuring that the public does not experience additional risks due to the implementation of new treatment package.

273. Second, monitoring is highly advisable to assess both the positive and the negative effects of the treatments. Before- and after-crash details are needed to develop a local set of CRFs and, in later years, to improve the accuracy of these CRFs. Regular (maybe annual) monitoring of the effectiveness of individual treatments allows for gradual refinements to the CRF for the treatment, consequently improving the accuracy of the economic assessments in future blackspot investigations.

274. Third, monitoring a treated site reinforces the need for strict protocols in crash investigation work. The monitoring process can easily be overlooked by a blackspot team as they conclude one investigation, hand it over for construction, and shift their focus to the next investigation. This oversight is understandable, as most blackspot teams focus on site treatment and crash reduction.

275. However, monitoring crash performance is necessary. It is crucial to plan for monitoring before implementing a treatment to ensure that adequate data (both before and after) are collected and clear objectives are set. Decisions should be made regarding the data to be collected before treatment and the time period during which data will be collected, allowing meaningful comparisons with posttreatment data.

276. The review period is important to ensure that the data are representative. Three years (sometimes longer) is a common time period for assessing treatment effects and determining the influence of the treatment on site crashes. This requires careful statistical analysis, accounting for external factors and noting that crash frequencies may be so low that any observed changes in crashes may not be statistically significant.

---

[8]   A common return of $4 in crash savings for every $1 spent on the site is quoted by one road authority (footnote 3).

"A blackspot program can yield a $4 benefit to the community for every $1 spent on treatments"

277.   A more detailed program evaluation, possibly undertaken by a university or a research group, will further investigate external factors that might influence crash rates at treated sites.

278.   The most common external factors that may need to be considered are the following:

- **Traffic flows.** Changes in traffic flows can affect crash rates, generally increasing with greater flows and decreasing with diminished flows. These traffic volume changes may result directly from the treatment itself or unrelated factors. Increases or decreases in crash rates may not follow a linear pattern.
- **National and local crash trends.** There may be more general trends in crashes happening over time in the region, such as the general reduction in crashes due to factors such as safer cars, overall road improvements, or rising fuel prices. These general changes should be accounted for, usually through the inclusion of similar control sites in the evaluation.

279.   The table of CRFs in Appendix 1 provides an example of the work conducted by a university research unit that regularly uses crash data from blackspot sites to evaluate the safety performance of applied treatments. The role of an independent and respected research unit enhances the credibility of any research. Independent research units, when engaged in the evaluation of a blackspot program, can be highly impactful on decision-makers when allocating blackspot funding for the upcoming year.

IX. References

1. African Development Bank. 2014. *Road Safety Manuals for Africa—Existing Roads: Reactive Approaches.* Côte d'Ivoire.

2. Alsop, P. and Longley, D. 2001 Under-Reporting of Motor Vehicle Traffic Crash Victims in New Zealand. *Accident Analysis and Prevention.* 33 (3).

3. Asian Development Bank (ADB). 2017. *Safely Connected: A Regional Road Safety Strategy For CAREC Countries, 2017–2030.* Manila.

4. ADB. 2018. *CAREC Road Safety Engineering Manual 1: Road Safety Audit.* Manila.

5. ADB. 2018. *CAREC Road Safety Engineering Manual 2: Safer Road Works.* Manila.

6. ADB. 2018. *CAREC Road Safety Engineering Manual 3: Roadside Hazard Management.* Manila.

7. ADB. 2021. *CAREC Road Safety Engineering Manual 4: Pedestrian Safety.* Manila.

8. ADB. 2022. *CAREC Road Safety Engineering Manual 5: Star Ratings for Road Safety Audit.* Manila.

9. Austroads. 2021. *Guide to Road Safety Part 2: Safe Roads.* AGRS02-21. Sydney.

10. Corben, B. et al. 1996. *Results of an Evaluation of TAC Funded Accident Black Spot Treatments.* Combined 18th Australian Road Research Board Transport Research Conference and Transit New Zealand Land Transport Symposium. 18 (5). pp. 343–359.

11. Federal Highways Administration, https://highways.dot.gov/safety

12. Federal Highways Administration. 2020. *The Safe System Approach.* Washington, DC.

13. Federal Highway Administration. 2022. Making our Roads Safer through a Safe System Approach. *Public Roads—Winter 2022.* 85 (4).

14. Global Road Safety Facility/World Bank. 2020. *Global Road Safety Facility: Leveraging Global Road Safety Successes.* Washington, DC.

15. Government of Pakistan, Ministry of Communications. 2019. *Guidelines for Road Safety Engineering (Part 1).* Islamabad.

16. McMahon, K. and Dahdah, S. 2008. The True Cost of Road Crashes: Valuing Life and the Cost of a Serious Injury. International Road Assessment Programme.

17. Nellthorp et al. 1998. Quoted in World Bank Note No TRN-16. *Valuation of Accident Reduction.*

18. Ogden, K. W. 1996. *Safer Roads: A Guide to Road Safety Engineering.* Avebury Technical, United Kingdom.

19. PIARC. 2017. *Vulnerable Road Users: Diagnosis of Design and Operational Safety Problems and Potential Countermeasures.*

20. TRACECA Technical Note 3. 2015. Regional Blackspot Management Guidelines.

21. United Nations Road Safety Collaboration group (UNRSC). 2020. *The Ten Step Plan for Safer Road Infrastructure.* Geneva.

22. Roads Corporation of Victoria. 2017. VicRoads Crash Reduction Factors for use in the Annual Works Program, Melbourne

23. World Bank. 2019. *Guide for Road Safety Opportunities and Challenges: Low- and Middle-Income Countries Country Profiles.* Washington, DC.

24. World Health Organization. 2023. *Global Status Report on Road Safety 2023.* Geneva.

25. World Bank 2021 Guide to Integrating Safety into Road Design. Washington, DC.

26. Wramborg, P. 2005 "A new approach to a safe and sustainable road structure and street design for urban areas." Road Safety on Four Continents Conference 2005, Warsaw, Poland.

Appendix 1: Crash Reduction Factors

No	Treatments	CRF % for All Casualty Crashes	CRF for Specific Casualty Crash Type	Treatment Life
	Crash Reduction Factors (CRFs) for VicRoads Road Safety Programs—as of July 2017			
	BRIDGE			
1	Bridge widening—one-lane two-way to two-lane two-way	30	90% head-on crashes	25
2	Bridge replacement—low standards to current standards	45		25
3	Bridge rail upgrade	20		20
4	Guard rail for bridge end post	20		20
5	Bridge warning signs	30		15
	DELINEATION (from no delineation, or where the existing delineation is no longer effective)			
1	Guideposts	7	28% dark crashes	15
2	Chevron alignment markers (CAMs)		30% Run off road (ROR) crashes on curves	15
3	CAMs—electronic		35% ROR crashes on curves	15
4	Raised reflectorized pavement markers (RRPM)	5		10
5	Edge lines—painted 100 millimeters (mm)	10		5
6	Centerlines—painted 100 mm	20		5
7	Centerline and edge line—painted 100 mm	25		5
8	Edge lines—tactile 100 mm	23		5
9	Centerlines—tactile 100 mm		84% fatal head-on crashes	5
10	Wider edge lines—painted 125 mm and 150 mm	37		5
11	Wide centerline (painted median) with tactile line markings, RRPMs, and chevrons on rural undivided roads with or without road widening. (This treatment may lead to provision of right turn lanes resulting in additional benefits that can offset the cost of road widening.)		80% head-on crashes 60% ROR crashes to left 58% serious injury crashes 90% fatal crashes	15 15 15 15
	INTERSECTION			
1	Grade separation of cross intersections	50		25
2	Roundabout (1 lane)—Urban environment Rural environment	70 80		25
3	Roundabout (2 lanes)—Urban environment Rural environment	60 70		25
4	Roundabout (3 lanes)—Urban environment Rural environment	50 60		25
5	Turbo roundabouts (2 lanes)—Urban environment Rural environment	70 80		25

continued on next page

Appendix 1: *continued*

No	Treatments	CRF % for All Casualty Crashes	CRF for Specific Casualty Crash Type	Treatment Life
6	Modify roundabout to achieve safe operating speeds		55% of crashes impacted by treatments, e.g., crashes on the modified leg(s)	25
7	Convert signals to roundabout (1 and 2 circulating lanes) Urban environment Rural environment	43 66		25
8	Staggered T (To be effective the intersection is currently ill-defined, average annual daily traffic [AADT] on major road < 2000 vehicles per day [VPD] and design of staggering distance >15 meters [m].)	40		25
9	Removal of Y-intersection by squaring the layout	85		25
10	Splitter islands urban	40		25
11	Splitter islands rural	35		25
12	Improve intersection definition by line marking	10		5
13	Sight distance improvement (remove obstruction)		55% reduction in crashes caused directly by obscured sightlines	25
14	Transverse rumble strips on approaches to rural intersections Speed limit 100 kilometers per hour (km/h) Speed limit 80 km/h	2.5 3		5
15	Banning movements (left, right, U-turns, etc.)		95% crashes of the affected Definition for classifying crashes (DCC)	25
16	New through lane (widen road)	20		25
17	New right turn lane—urban	35		25
18	New left turn lane—urban	30		25
19	Lengthen turning lane—urban	15		25
20	Widen road (rural) for (large radius) turn lane—protected Large radius = left turns for driving on right side of road		70% crashes DCC 130, 132, 133, 134, 136	25
21	Widen road (rural) for large radius turn bypass		50% crashes DCC 130, 132, 133, 134, 136	25
22	Widen road (rural) for (small radius) turn lane—protected Small radius = right turns for driving on the right side of road		90% crashes DCC 130, 131, 133, 135, 137 50% crashes DCC 110, 113, 116, 122, 123	25
23	No sign to stop sign at T-intersection	20		15
24	No sign to stop sign at X-intersection	35		15
25	No sign to give way sign	15		15
26	Upgrade form give way sign to stop sign	5		15
27	Advance intersection warning sign—static	15		15

continued on next page

Appendix 1: *continued*

No	Treatments	CRF % for All Casualty Crashes	CRF for Specific Casualty Crash Type	Treatment Life
28	Vehicle activated warning signs—cross intersection	50		15
29	Vehicle activated warning signs—T-intersection	75		15
30	Variable message sign	20		15
31	Advisory speed sign	25		15
32	Speed limit reduction from 60 to 40 km/h through the intersection	15		15
33	Skid resistance overlay (high passenger service vehicle [PSV])	20	40% wet crashes	5
34	Skid resistance overlay (calcined bauxite)	40	50% wet crashes	10
35	From no to new street lighting—intersections		50% dark crashes	20
	INTERSECTION—TRAFFIC SIGNALS			
1	New traffic signals	45	60% serious casualty crashes	15
2	Improve signal visibility (mast arms, etc.)	25		15
3	Signal remodel (phasing, controller, lanterns, etc.)	10		15
4	Increase all red interval (approx. 1 - 2 sec)	14		15
5	Fixed digital red light cameras for existing signalized intersections	14	All injury crashes	10
6	Fixed digital speed and red light cameras	26	Crashes from all approaches 44% crashes from the targeted approach	10
7	Upgrade from existing red light cameras to red light and speed cameras	19 72 41 20	All casualty crashes Serious injury crashes Right angle/right turn through crashes Rear-end crashes	10
8	Ban large radius turn at signalized intersection	95		25
9	Fully controlled large radius turn—from no control or partially controlled turn		80% crashes DCC 121 and U-turn during turns	15
10	Partially controlled left/right turn—from no control	4	14% crashes DCC 121 during turn periods	15
11	Signalization of left/right turn lane	30	70% crashes DCC 106, 116, 123, 131, 135, 137 plus crashes occurred in the turn lane only	15
12	Advance warning flashers	18		15
13	Linking existing signals	15		15
	MOTORCYCLISTS			
1	Sealing bell mouths		20% motorcycle (M/C) crashes only	25

continued on next page

Appendix 1: *continued*

No	Treatments	CRF % for All Casualty Crashes	CRF for Specific Casualty Crash Type	Treatment Life
2	Rub rail (underneath the W beam railing)		15% M/C crashes only	20
3	Modular impact cushion for small signposts (warning signs, etc.)		15% M/C crashes only	20
	PAVEMENT IMPROVEMENTS			
1	Road reconstruction	25		25
2	Duplication short length	15	95% overtaking related crashes	25
3	Raised median—urban	45		25
4	Raised median—rural	55		25
5	Painted median	20	50% crashes DCC 120, 100, 101, 102	5
6	Wide centerline (painted median) with tactile line markings, RRPMs, and chevrons on rural undivided roads with or without road widening. (This treatment may lead to provision of right turn lanes resulting in additional benefits that should be part of safety gains.)		80% head-on crashes 60% ROR crashes to left 58% serious injury crashes 90% fatal crashes	10 10 10 10
7	Widen pavement to standard lane widths	10		25
8	Overtaking lane	23	95% overtaking related crashes	25
9	Add extra lane—urban	30	If there is NO increase in operating speed or speed limit	25
10	Add extra lane—rural	35	If there is NO increase in operating speed or speed limit	25
11	Seal shoulder and edge line from nothing		30% ROR crashes 14% fatal and serious injury (FSI) crashes	25
12	Seal shoulder and tactile edge line from nothing		35% ROR crashes	25
13	Skid resistance overlay (high PSV)	20	40% wet crashes	5
14	Skid resistance overlay (calcined bauxite)	40	50% wet crashes	10
15	Curve realignment		40% crashes on curves	25
16	Correct superelevation	20		25
17	Side slope flattening—general	6		25
18	Flatten from 2:1 to 4:1	6		25
19	Flatten from 2:1 to 5:1	9		25
20	Flatten from 2:1 to 6:1	12		25
21	Flatten from 2:1 to 7:1 or flatter	15		25
22	Flatten from 4:1 to 5:1	3		25
23	Flatten from 4:1 to 6:1	7		25
24	Flatten from 4:1 to 7:1 or flatter	11		25

continued on next page

Appendix 1: *continued*

No	Treatments	CRF % for All Casualty Crashes	CRF for Specific Casualty Crash Type	Treatment Life
	PEDESTRIANS & CYCLISTS			
1	Pedestrian refuges		50% pedestrian crashes 67% fatal pedestrian crashes	25
2	Painted (flush) median—urban		50% pedestrian crashes	5
3	Pedestrian operated signals		39% pedestrian Crashes	15
4	Exclusive pedestrian signal phase at intersection		50% pedestrian crashes	15
5	Improved signal timing		35% pedestrian crashes	15
6	Upgrade marked crossing (non-signalized) to signalized crossing		27% pedestrian crashes	15
7	Flashing "give way to pedestrians" sign plus pedestrian phase priority at signalized intersection		35% pedestrian crashes	15
8	Pedestrian overpass		85% pedestrian crashes	25
9	Pedestrian fencing and barriers		23% pedestrian crashes	20
10	Improved lighting at pedestrian crossings		60% pedestrian crashes nighttime only 30% pedestrian crashes	20
11	Dwell on red signal phase	47	50% pedestrian crashes during the operating time	15
12	On road bicycle paths		30% bicycle crashes	5
13	On road bicycle paths—green marking	25		5
14	40 km/h speed reduction along a shopping streets with electronic and static signs	8	15% pedestrian crashes	15
15	50 km/h default speed limit		23% pedestrian crashes 41% pedestrian FSI crashes	20
16	50 km/h default speed limit reduced from 70km/h		61% pedestrian crashes	20
17	Raised wombat crossing	63	73% pedestrian crashes	20
18	Ban parking		30% pedestrian crashes	20
19	Barnes dance crossing		9% pedestrian crashes	20

continued on next page

Appendix 1: *continued*

No	Treatments	CRF % for All Casualty Crashes	CRF for Specific Casualty Crash Type	Treatment Life
20	Footpath and shoulder provision		88% pedestrian crashes (walking along)	25
21	High visibility crosswalk (zebra crossings with additional markings, lighting, colors, etc.)		44% pedestrian crashes	20
22	Traffic signals (new control at intersections)		30% pedestrian crashes	15
23	Puffin crossings		26% pedestrian crashes	20
24	Raised intersection platforms	20	8% pedestrian crashes	20
25	Traffic calming—all environment	20		20
26	Traffic calming and 30 km/h speed limit		65% pedestrian crashes 50% ped FSI crashes	20
	RAIL CROSSING AT GRADE			
1	From nothing to signage		25% train crashes only	15
2	From signage to light and bells		50% train crashes only	15
3	From lights and bells to barriers		45% train crashes only	15
4	From signage to barriers		67% train crashes only	15
5	Improved sight distance		44% train crashes only	15
6	Rural rail crossing—speed limit 100 km/h	2		5
7	Rural rail crossing—speed limit 80 km/h	2.5		5
	ROADSIDE HAZARDS—RUN-OF-ROAD CRASHES			
1	10%–20%		20% ROR crashes	20
2	21%–40%		25% ROR crashes	20
3	41%–60%		30% ROR crashes	20
4	60%–70%		35% ROR crashes	20
5	71%–80%		40% ROR crashes	20
6	Replace rigid poles with impact absorbent poles		25% ROR crashes	20
7	Replace rigid poles with slip-based poles		35% ROR crashes	20
8	Embankment treatment		40% ROR crashes	25
	ROADSIDE HAZARDS—BARRIERS & CUSHIONS **Rigid barriers (concrete) are not recommended**			
1	Wire rope safety barrier (WRSB) including on median	76	32% FSI crashes (100 km/h) 61% FSI crashes (110 km/h) 85% ROR crashes	20

continued on next page

Appendix 1: *continued*

No	Treatments	CRF % for All Casualty Crashes	CRF for Specific Casualty Crash Type	Treatment Life
2	WRSB on median of divided roads		14% FSI crashes (100 km/h) 26% FSI crashes (110 km/h) 81% other injury crashes ROR to right 98% serious injury crashes ROR to right	20
3	Centerline wire rope safety barriers—undivided rural	13	46% ROR and head-on crashes	20
4	W-beam safety barrier (guardrail) including on median	40	32% FSI crashes (100 km/h) 61% FSI crashes (110 km/h) 45% ROR crashes	20
5	Other flexible barrier (Ezy-Guard Smart)		18% FSI crashes 65% ROR crashes	20
6	Guard rail at culvert		25% ROR crashes	20
7	Impact attenuator		20% ROR crashes	20
8	Crash cushions for poles/trees (e.g., raptor)		50% ROR hit objects serious injury crashes	20
9	Rub rail below barrier		15% motorcycle crashes only	
10	Modular impact cushion for small posts (CAMS, regulatory signs)		15% motorcycle crashes only	
	SIGNAGE—STATIC SIGNS			
1	Guidance sign (directional, information, etc.)	15		15
2	Advisory speed sign	25		15
3	Curve warning sign	25		15
4	Advance warning sign—static	15		15
	SIGNAGE—VEHICLE ACTIVATED SIGNS (VAS)			
1	Speed roundels (speed limit + slow down)	39		15
2	Curve warning signs—winding road + too fast	53		15
3	Curve warning signs—right sharp turn + too fast	67		15
4	Curve warning signs—right curve + slow down	40		15
5	Curve warning signs—left curve + slow down	23		15
6	Intersection warning signs—cross intersection	50		15
7	Intersection warning signs—T-intersection	75		15
8	Variable message signs	20		15
	SPEED LIMIT REDUCTION—STATIC SIGNS			
1	From 100 to 90 km/h		20% FSI crashes	15
2	From 100 to 80 km/h	15		15
3	From 80 to 60 km/h	20		15

continued on next page

Appendix 1: *continued*

No	Treatments	CRF % for All Casualty Crashes	CRF for Specific Casualty Crash Type	Treatment Life
4	From 60 to 50 km/h	20		15
5	All other reductions in speed limit	15		15
6	From 70 to 50 km/h		61% pedestrian crashes	15
7	50 km/h default speed limit		23% pedestrian crashes	15
	SPEED LIMIT REDUCTION—VAS			
1	On major road when activated by vehicles on minor roads from 100 to 70 km/h (rural)	91	95% FSI crashes	15
	SPEED LIMIT REDUCTION—GATEWAY TREATMENT FOR RURAL TOWNS/VILLAGES (SIGNS AND PINCH POINT)			
1	Reduce from 100 to 80 km/h	35	41% FSI crashes	15
	SPEED REDUCTION—RAISED SAFETY PLATFORMS			
1	Raised safety platforms (mid-block)	47		20
2	Raised safety platforms on approaches to signalized intersections	40	8% pedestrian crashes	20
3	Raised safety platforms on approaches at priority intersections	35	8% pedestrian crashes	20
4	Raised intersections	30		20
	SPEED REDUCTION			
1	Traffic calming—all environment (with 30 km/h speed limit)	20	50% FSI pedestrian crashes 65% pedestrian crashes	20
2	Dragon teeth pavement marking on approach to intersections or rural curves—100 km/h zone		10% speed-related crashes	5
3	Dragon teeth pavement marking on approach to intersections or rural curves—80 km/h zone		21% speed-related crashes	5
	SPEED ENFORCEMENT			
1	Mobile camera—urban and rural	20		15
2	Fixed camera—urban and rural	30		15
	STREET LIGHTING (CURRENTLY NONEXISTENT)			
1	New street lighting—intersections		50% dark crashes only	20
2	New street lighting—mid-blocks		40% dark crashes only	20
3	New street lighting—motorway/freeway interchanges		50% dark crashes only	20
4	New street lighting—railway crossings		60% dark crashes only	20

continued on next page

Appendix 1: *continued*

No	Treatments	CRF % for All Casualty Crashes	CRF for Specific Casualty Crash Type	Treatment Life
	STREET LIGHTING (CURRENTLY EXISTING)			
1	Upgrading existing lighting		35% dark crashes only	20
2	Upgrading lighting at pedestrian crossings		60% pedestrian crashes nighttime only	20

Source: Roads Corporation of Victoria. 2017. VicRoads Crash Reduction Factors for use in the Annual Works Program, Melbourne.

How to use the CRFs

A. The principle is to apply the CRF to the crashes relevant to the proposed treatments.

B. For multiple treatments at intersections or very short lengths, DO NOT SUM the various CRFs. Select one of the methods shown below to estimate an overall project CRF, or seek advice from the head office if unsure:

   1. Use the Project Submission Pro-Forma to estimate the crash savings for each treatment using targeted crashes (maximum three key treatments) with specific CRFs. Then, divide the sum of [Predicted No. of All Casualty Crashes Saved per year] by the sum of [No. of All Casualty Crashes per year] to obtain the overall CRF for the project.
   2. Use the formula $CRFt = 0.66 \times [1-(1-CRF1)(1-CRF2)(1-CRF3)]$ if each of the three key treatments addresses all crashes that occurred at the intersection. Note that this method requires approval from the head office team.

C. If a project contains one major treatment and several minor treatments, adopt only the CRF for the major treatment that can address the majority of the crashes.

D. When a second project is developed for the same site for which a bid has been submitted or a project has been funded, the CRF for the treatments of the second project is calculated using crashes and CRFs corresponding to the targeted crashes that remain to be addressed.

E. For special cases where the value of CRF is expected to be higher than the above estimate or for innovative treatments not listed here, please contact head office.

Descriptions for Classifying Crashes (DCC)

Pedestrian on foot in toy/pram	Vehicles from adjacent directions (intersections only)	Vehicles from opposing directions	Vehicles from same direction	Maneuvering	Overtaking	On path	Off path on straight	Off path on curve	Passenger and miscellaneous
100 NEAR SIDE	110 CROSS TRAFFIC	120 HEAD ON (NOT OVERTAKING)	130 REAR END	140 U TURN	150 HEAD ON (INCLUDING SIDE SWIPE)	160 PARKED	170 OFF CARRIAGEWAY TO LEFT	180 OFF CARRIAGEWAY ON RIGHT BEND	190 FELL IN/FROM VEHICLE
101 EMERGING	111 THRU-RIGHT	121 RIGHT-OPPOSING THRU	131 LEFT REAR END	141 U TURN INTO FIXED OBJECT/PARKED VEHICLE	151 OUT OF CONTROL	161 DOUBLE PARKED	171 LEFT OFF CARRIAGEWAY INTO OBJECT/PARKED VEHICLE	181 OFF RIGHT BEND INTO OBJECT/PARKED VEHICLE	191 LOAD OR MISSILE STRUCK VEHICLE
102 FAR SIDE	112 THRU-LEFT	122 LEFT-OPPOSING THRU	132 RIGHT REAR END	142 LEAVING PARKING	152 PULLING OUT	162 ACCIDENT OR BROKEN DOWN	172 OFF CARRIAGEWAY TO RIGHT	182 OFF CARRIAGEWAY ON LEFT BEND	192 STRUCK TRAIN
103 PLAYING, WORKING, LYING, STANDING ON ROAD	113 RIGHT-THRU	123 RIGHT-OPPOSING LEFT	133 LANE SIDE SWIPE	143 ENTERING PARKING	153 CUTTING IN	163 VEHICLE DOOR	173 RIGHT OFF CARRIAGEWAY INTO OBJECT/PARKED VEHICLE	183 OFF LEFT BEND INTO OBJECT/PARKED VEHICLE	193 STRUCK RAILWAY CROSSING EQUIPMENT
104 WALKING WITH TRAFFIC	114 TWO RIGHT TURNS	124 RIGHT-OPPOSING RIGHT	134 LANE CHANGE RIGHT	144 PARKING VEHICLES ONLY	154 PULLING OUT-REAR END	164 PERMANENT OBSTRUCTION ON CARRIAGEWAY	174 OUT OF CONTROL ON CARRIAGEWAY	184 OUT OF CONTROL ON CARRIAGEWAY	194 PARKED CAR RUN AWAY
105 FACING TRAFFIC	115 RIGHT-LEFT	125 LEFT-OPPOSING LEFT	135 LANE CHANGE LEFT	145 REVERSING		165 TEMPORARY ROADWORKS	175 OFF END OF ROAD/T INTERSECTION		
106 ON SIDEWALK/MEDIAN	116 LEFT-THRU		136 RIGHT TURN SIDE SWIPE	146 REVERSING INTO FIXED OBJECT/PARKED VEHICLE		166 STRUCK OBJECT ON CARRIAGEWAY			
107 DRIVEWAY	117 LEFT-RIGHT		137 LEFT TURN SIDE SWIPE	147 EMERGING FROM DRIVEWAY/LANE		167 STRUCK ANIMAL (NOT RIDDEN)			198 OTHER
108 STRUCK WHILE BOARDING OR ALIGHTING VEHICLE	118 TWO LEFT TURNS			148 FROM SIDEWALK					
109 OTHER PEDESTRIAN	119 OTHER ADJACENT	129 OTHER CROSSING	139 OTHER SAME DIRECTION	149 OTHER MANEUVERING	159 OTHER OVERTAKING	169 OTHER ON PATH	179 OTHER OFF PATH ON STRAIGHT	189 OTHER OFF PATH ON CURVE	199 UNKNOWN

1. Traffic accident descriptions should be determined by first selecting a column using the text above each column and then by appropriate diagram.
2. The diagram chosen should best describe the general movement of vehicles involved in the initial event. It does not assign a cause to the accident.
3. Supplementary codes have been defined for most diagrams. These codes give further detail of the initial event.
4. The letters a & b identify individual vehicles involved when the traffic accident description is linked with other vehicle/driver information.

Note: These charts are valuable tools for helping police officers complete crash report forms. They provide a specific code for almost every type of crash.
Source: Roads Corporation of Victoria. 2017. Victoria's Road Safety Program for 2017, Appendix A VicRoads Crash Reduction Factors for use in the Annual Works Program, Melbourne, Australia.

1. REPORT NO. 3. POLICE STATION	TRAFFIC ACCIDENT REPORT FORM	2. PROVINCIAL OFFICE 4. REGIONAL OFFICE

5. NUMBER OF VEHICLES INVOLVED			9. ACCIDENT SEVERITY		10. Month 11. Day 12. Year
6. NUMBER OF DRIVER CASUALTIES			F. Fatal Accident	DATE	
7. NUMBER OF PASSENGER CASUALTIES			S. Serious Injury Accident M. Minor Injury Accident	13. DAY OF WEEK	
8. NUMBER OF PEDESTRIAN CASUALTIES			D. Property Damage Only	14. TIME (Military Time)	

15. JUNCTION TYPE
1. Not at Junction   5. Y
2. +   6. (roundabout symbol)
3. T   7. Railway
4. ⊥   8. Other

16. TRAFFIC CONTROL
1. None
2. Centerline
3. Pedestrian Crossing
4. School Crossing
5. Police Controlled
6. Traffic Lights
7. Stop Sign
8. Give Way
9. Other ......

17. COLLISION TYPE
1. Head On
2. Rear End
3. Right Angle
4. Side Swipe
5. Overturned Vehicle
6. Hit Object in Road
7. Hit Object off Road
8. Hit Parked Vehicle
9. Hit Pedestrian
10. Hit Animal
11. Other ....

17. MOVEMENT
1. 1-Way
2. 2-Way

19. SEPARATION
1. Median
2. No Median

20. WEATHER	21. LIGHT	22. ROAD CHARACTER	23. SURFACE CONDITION	24. SURFACE TYPE	25. MAIN CAUSE	26. ROAD CLASS
1. Fair 2. Rain 3. Wind 4. Smoke 5. Fog 6. Dazzle 7. Storm	1. Daylight 2. Dawn/Dusk 3. Night (lit) 4. Night (unit)	1. Straight + Flat 2. Curve only 3. Incline only 4. Curve + Incline 5. Bridge ...... 6. Crest	1. Dry 2. Wet 3. Muddy 4. Flooded 5. Other	1. Concrete 2. Asphalt 3. Gravel 4. Earth	1. Vehicle Defect 2. Road Defect 3. Human Error 4. Other	1. National 2. Provincial 3. City 4. Municipal 5. Barangay

27. ROAD REPAIRS 1. Yes 2. No	28. HIT & RUN 1. Yes 2. No	29. LOCATION TYPE 1. Urban Area 2. Rural Area

LOCATION

Name of City/Town .... Distance .... (km/m)

Name of Road .... Between — Landmark 1 ....   Distance .... (km/m)
Landmark 2 ....   Distance .... (km/m)

JUNCTION ACCIDENT ONLY Name of SECOND Road ....   Distance .... (km/m)

LOCATION SKETCH MAP Show site in relation to prominent landmarks such as KM posts or Major intersections. Mark distances to the landmarks

COLLISION DIAGRAM SKETCH Mark the position and direction of each vehicle and details of the road layout at the site of the accident

Signatures: Driver 1 ....   Driver 2 ....

POLICE DESCRIPTION OF ACCIDENT
....
....
....
....
....
....
....

WITNESSES
1. Name ....
   Address ....
2. Name ....
   Address ....

INVESTIGATING OFFICER

Name/Rank ....   Date ....

SUPERVISING OFFICER

Name/Rank ....   Date ....

DRIVER STATEMENTS
Driver 1 ....
....
Driver 2 ....
....

ACTION TAKEN

RECOMMENDATION

STATUS OF CASE

Additional form(s) will be needed if there are more than 2 vehicles, more than 4 passenger casualties or more than 2 pedestrian casualties.
Fill in the report no, provincial office, police station and dates and fix forms together securely.

1. REP NO.	2. PROV OFFICE	3. POL STN	4. REG OFFICE	DATE

VEHICLE 1	30. VEHICLE PLATE NUMBER	DRIVER 1	NAME

31. OWNER'S NAME & ADDRESS — ADDRESS

CHASSIS NUMBER — **32. ENGINE NUMBER** — LICENSE NUMBER

33. INSURANCE — OR/CR DETAILS — LICENSE TYPE — EXPIRY DATE

MANUFACTURER (MAKE) — MODEL/YEAR — **40. DRIVER SEX** — **42. DRIVER INJURY**
1. Fatal   3. Minor
2. Serious 4. Not Injured
Hospital:

34. VEHICLE TYPE
1. Bicycle   7. Bus
2. Pedicab   8. Truck (Rigid)
3. Motor Cycle 9. Truck (Artic)
4. Tricycle   10. Van
5. Car   11. Animal
6. Jeepney   12. Other
....

35. VEHICLE MANEUVER
1. Left Turn   7. Overtaking   13. Parked On Road
2. Right Turn   8. Going Ahead   14. Other
3. U Turn   9. Reversing   ....
4. Cross Traffic 10. Sudden Start
5. Merging   11. Sudden Stop
6. Diverging   12. Parked Off Road

41. DRIVER AGE

43. DRIVER ERROR
1. None   6. No Signal
2. Fatigued/Asleep 7. Bad Overtaking
3. Inattentive   8. Bad Turning
4. Too Fast   9. Using Cell Phone
5. Too Close   10. Other
....

36. LOADING
1. Legal
2. Over Loaded
3. Unsafe Load

37. DIRECTION
1. North
2. South
3. East
4. West

38. VEHICLE DEFECT
1. None   5. Tires
2. Lights 6. Multiple
3. Brakes 7. Other
4. Steering ....

39. VEHICLE DAMAGE
1. None   5. Left
2. Front   6. Roof
3. Rear   7. Multiple
4. Right   8. Other
....

44. ALCOHOL/DRUGS
1. Alcohol Suspected
   Drugs Suspected
2. Not Suspected

45. SEAT BELT/HELMET
1. Seat Belt/Helmet Worn
2. Not Worn
3. Not Worn Correctly

VEHICLE 2	30. VEHICLE PLATE NUMBER	DRIVER 2	Name

31. OWNER's NAME & ADDRESS — ADDRESS

CHASSIS NUMBER — **32. ENGINE NUMBER** — LICENSE NUMBER

33. INSURANCE — OR/CR DETAILS — LICENSE TYPE — EXPIRY DATE

MANUFACTURER (MAKE) — MODEL/YEAR — **40. DRIVER SEX** — **42. DRIVER INJURY**
1. Fatal   3. Minor
2. Serious 4. Not Injured
Hospital:

34. VEHICLE TYPE
1. Bicycle   7. Bus
2. Pedicab   8. Truck (Rigid)
3. Motor Cycle 9. Truck (Artic)
4. Tricycle   10. Van
5. Car   11. Animal
6. Jeepney   12. Other
....

35. VEHICLE MANEUVER
1. Left Turn   7. Overtaking   13. Parked On Road
2. Right Turn   8. Going Ahead   14. Other
3. U Turn   9. Reversing   ....
4. Cross Traffic 10. Sudden Start
5. Merging   11. Sudden Stop
6. Diverging   12. Parked Off Road

41. DRIVER AGE

43. DRIVER ERROR
1. None   6. No Signal
2. Fatigued/Asleep 7. Bad Overtaking
3. Inattentive   8. Bad Turning
4. Too Fast   9. Using Cell Phone
5. Too Close   10. Other
....

36. LOADING
1. Legal
2. Over Loaded
3. Unsafe Load

37. DIRECTION
1. North
2. South
3. East
4. West

38. VEHICLE DEFECT
1. None   5. Tires
2. Lights 6. Multiple
3. Brakes 7. Other
4. Steering ....

39. VEHICLE DAMAGE
1. None   5. Left
2. Front   6. Roof
3. Rear   7. Multiple
4. Right   8. Other
....

44. ALCOHOL/DRUGS
1. Alcohol Suspected
   Drugs Suspected
2. Not Suspected

45. SEAT BELT/HELMET
1. Seat Belt/Helmet Worn
2. Not Worn
3. Not Worn Correctly

PASSENGER CASUALTIES   Complete 1 FULL line for each passenger casualty   *= See Reference boxes below

NAME & ADDRESS	46. VEH. NO	47. SEX	48. AGE	49. * INJURY/HOSP	50. * POSITION	51. * ACTION
1.						
2.						
3.						
4.						

PEDESTRIAN CASUALTIES   Complete 1 FULL line for each pedestrian casualty   *= See Reference boxes below

NAME & ADDRESS	52. SEX	53. AGE	54. * INJURY/HOSP	55. * POSITION	56. * ACTION
1.					
2.					

FOR REFERENCE ONLY DO NOT CIRCLE	49. PASSENGER INJURY 54. PEDESTRIAN INJURY	50. PASSENGER POSITION	51. PASSENGER ACTION	55. PEDESTRIAN LOCATION	56. PEDESTRIAN ACTION
	F. Fatal S. Serious M. Minor	1. Front Seat 2. Rear Seat 3. M/Cycle Passenger 4. Bus Passenger 5. Outside—Sitting 6. Outside—Standing	1. None 2. Boarding 3. Alighting 4. Falling 5. Other	1. On Pedestrian Crossing 2. Within 50 m Ped. Crossing 3. On Central Refuge 4. In Road Centre 5. On Footpath/Verge	1. None 2. Crossing Road 3. Walking along Road 4. Walking along Edge 5. Playing on Road 6. On Footpath

Glossary

Active crossing. A pedestrian facility that grants right of way to pedestrians only when activated (a red signal is usually activated to instruct drivers to stop)

Advance warning sign. A sign placed in advance of a hazard to provide early warning to approaching traffic

Blacklength. A road segment (usually 1 kilometer or longer) with a high number of casualty crashes

Blackspot. A location on the road network with a high number of casualty crashes

Benefit–cost ratio. The ratio of the benefit of an action divided by the cost of that action (a ratio above 1.0 indicates a positive outcome)

CAREC highway. One of the designated national/international highways under the Central Asia Regional Economic Cooperation (CAREC) program

Casualty. A serious injury or a death

Casualty crash. A road crash in which at least one person is killed or seriously injured

Client. The road authority responsible for the road/highway

Collision. Synonymous with crash (see below)

Consultant. The client's representative for the project

Contractor. The company contracted to undertake the work for the client

Crash. A rare, random multifactorial event in which one or more road users fails to cope with their environment

Crash reduction factor. The percentage reduction in casualty crashes expected when a given safety treatment is installed (based on before- and after-crash studies at various sites)

Curb extension. A change in the curb line that extends into the road, decreasing the road's effective width, thereby improving pedestrian safety by reducing exposure to traffic and enhancing pedestrian visibility before crossing

Curb ramp. A smooth access point between the footpath and the road (while available for all pedestrians, it is particularly beneficial for those with mobility restrictions, such as individuals using wheelchairs, strollers, walkers, or handcarts; pedestrians can use the ramp to move between the road and footpath without having to step onto and off high curbstones)

Delineation. A general term for signs and devices that clearly define the designated traffic path

Fatal crash. A road crash in which at least one person is killed or dies within 30 days

Fatality. A death caused by a road crash

Footbridge. A structure enabling pedestrians to cross over a road, providing spatial separation; synonymous with pedestrian overpass

Frangible. The ability of a device, including structure supports, posts, and poles, to break away or deform upon impact by an errant vehicle without posing a significant risk of serious injury to vehicle occupants

Gateway. The generic term used for an entrance (made of signs, pavement markings, and gantries) on the approach to a town or village that welcomes drivers and helps inform them they are entering a different driving environment

High-speed road. A road where vehicle speeds are typically greater than 60 kilometers per hour (km/h)

Intoxicated road user. An individual whose perception, mood, thinking processes, and motor skills are affected by drugs, commonly associated with alcohol but can also involve other substances that impact the central nervous system

Low-speed road. A road where vehicle speeds are typically 60 km/h or less

Multilane. Two or more traffic lanes in one direction

Passive crossing. A pedestrian facility that requires drivers to yield the right of way to pedestrians on the crossing without the need for pedestrian operated signals (the pedestrian or zebra crossing is a common example)

Pedestrian. Any person on foot, including those in prams, pushers, wheelchairs; on skateboards; and walking with a bicycle

Pedestrian crossing. A passive crossing with painted stripes on the road and regulatory signs nearby facing the drivers (synonymous zebra crossing)

Pedestrian operated signal. A traffic signal that includes push buttons for pedestrians to record their intention to cross, often found at intersections or mid-block locations, creating traffic gaps to give pedestrians time separation

Pedestrian overpass. Synonymous with a footbridge (see above)

Pedestrian underpass. A grade-separated facility allowing pedestrians to walk beneath a road, serving as a form of spatial separation (aka subway)

Posted speed limit. The signed maximum legal speed limit for a road

Property damage. A crash in which no one is injured, but property (vehicles and/or other property) is damaged

PUFFIN crossing. A version of a pedestrian operated signal with an overhead microwave detector to determine if a slow-moving pedestrian requires additional clearance time (name derived from pedestrian user-friendly intelligent crossing)

Refuge. A physical island in the center of a road providing space for pedestrians to stand while staging their road crossing, making it a form of spatial separation

Roadside. The area between the road reservation boundary and the edge of the shoulder or traffic lane in the absence of a shoulder (the median between carriageways of a divided road is also a part of the roadside)

Roadside hazard. Any feature located in the clear zone (along the roadside or within the median) that could cause significant injury to the occupants of an errant vehicle (CAREC Road Safety Engineering Manual 3 [2018] explains this topic in detail)

Road user. Any driver, rider, passenger, or pedestrian using the road

Roadway. That portion of the road for the use of vehicles, including the shoulders and auxiliary lanes

Road work. Any work on a road or a roadside that has the potential to disturb traffic flow and/or safety (CAREC Road Safety Engineering Manual 2[2018] explains this topic in detail)

Safe System. A holistic approach to road safety that involves the interactions among roads and roadsides, travel speeds, vehicles, and road users, recognizing human error and aiming to prevent death or serious injuries resulting from crashes by prioritizing forgiveness in the system

Safety barrier. A physical barrier separating a hazard from the traveled way, designed to resist penetration by an out-of-control vehicle and (as far as practicable) to redirect the colliding vehicle back into the traveled path

Senior citizen. An individual who is 65 years of age or older

Serious injury. An injury that causes the person to spend at least one night in hospital for treatment

Slight injury. A low level of injury that requires medical attention but not a hospital stay

Shared zone. A length of road or street that is signed to create a zone in which pedestrians have priority over the motor vehicles, where motor vehicles are restricted in speed (usually 10 or 20 km/h) and where vehicle parking is permitted only in marked bays

Subway. Synonymous with pedestrian underpass (see above)

Traffic. All vehicles (including cars, trucks, buses, bicycles, motorcycles, and animal-drawn vehicles), persons, and animals traveling on the road

Traffic calming. The use of road humps, chicanes, roundabouts, pavement textures and markings, and other physical devices to reduce vehicle speeds in local streets

Traffic control devices. The signs, signals, crossings, barriers, and other devices placed on or near the road to regulate, warn, or guide road users

Traffic lane. A portion of a road used for the movement of traffic (does not include shoulders)

Two-way roadway. A roadway with lanes allotted for use by traffic in opposing directions without physical separation between them

Vulnerable road user. A road user group considered vulnerable, due to their relative frailty in the event of a collision with a motor vehicle (e.g., pedestrians, bicyclists, motorcyclists, and animal-drawn vehicles/carts)

Zebra crossing. Synonymous with pedestrian crossing (see above)

www.ingramcontent.com/pod-product-compliance
Lightning Source LLC
Chambersburg PA
CBHW050049220326
41599CB00045B/7339